FOREST
ON ZONES
CIAL LIMIT
Lake Superior
NORTHERN
SIN
Lake Michigan
ntario
MADISON
MILWAUKEE
FOREST
CHICAGO
GARY
LINOIS
PEORIA
PENINSULA
INDIAN
PRAIRIE
BEECH
SPRINGFIELD
Wabash R.
EVANSVILLE
SBURGH
250
THE MIDDLE WEST
Allen Carroll

The Shaping of America's Heartland

The Naturalist's America

Houghton Mifflin Company Boston 1977

The Shaping of America's Heartland

The Landscape of the Middle West

by Betty Flanders Thomson

Illustrated with Photographs and Maps

THE NATURALIST'S AMERICA

1. *The Appalachians* by Maurice Brooks
2. *Sierra Nevada* by Verna R. Johnston
3. *Desert* by Ruth Kirk
4. *The Shaping of America's Heartland* by Betty Flanders Thomson

Maps are by Allen Carroll

The diagram on page 61, from "Peat Samples for Radiocarbon Analysis," American Journal of Science 249: 473–511, 1951, is reproduced by permission of the authors, E. J. Deevey, Jr., and J. E. Potzger.

Library of Congress Cataloging in Publication Data

Thomson, Betty Flanders, date
The shaping of America's heartland.

(The Naturalist's America)
Bibliography: p.
Includes index.
1. Landforms—Middle West. 2. Botany—Middle West—Ecology. 3. Hydrology—Middle West. I. Title.
GB428.M52T46 551.4'0977 77-9325
ISBN 0-395-24760-8

Printed in the United States of America

V 10 9 8 7 6 5 4 3 2 1

To George and Virginia Avery —
for encouragement, forbearance, and long friendship's sake

Editor's Preface

In this day of environmental awareness the nature-oriented traveler is taking a fresh look at the landscape. The serious bird watcher is no longer content with merely ticking off the warblers, hawks, and waterfowl on his little white checklists; nor is the botanist, the lepidopterist, or the herpetologist satisfied only with recording new stations for rare species. They want to know why the animals and plants are where they are and how they fit into the ecosystem. Basic to this understanding is the geological and climatic history of the land and an awareness of how man has modified things.

It is not surprising that so many ecologists have concentrated their attentions on the mountain states. The contrasts are greater there; the interrelationships between animals and their environment are more obvious. Birds and plants are distributed altitudinally, changing dramatically as one ascends the slopes, as Clinton Hart Merriam noted many years ago when he formulated his life-zone concept.

Not so clearly defined are the dynamics of the broad lands that stretch between the Appalachians and the Rockies, although there is a gradual south-to-north transition somewhat comparable to the altitudinal effect of the mountains. It is this vast, relatively flat, sometimes gently rolling intermontane area, bordered by the sultry south and the cool Canadian north, that Betty Flanders Thomson interprets for us in this very readable book. Popularly known as the Middle West or Midwest, it blends imperceptibly somewhere short of the hundredth meridian with the Far West. Prairie farms give way to drier, more open plains, a sort of "twilight zone" where eastern birds reach their westernmost limits along the river valleys and western birds penetrate eastward along the relatively arid ridges.

Many readers, I am sure, habitually use one or more of the *Field Guides.* Miss Thomson's book, although not a field guide, can be read as a companion volume to the series. It will be illuminating to anyone who lives in or travels through the central parts of the United States.

One does not have to be a naturalist to observe and interpret the countryside, but it helps. Certainly no one can totally ignore the landscape, even though they may drive through it as fast as the law permits. They may be diverted by billboards, or they may find the featureless cornfields of Illinois or Iowa boring, but the landscape is inexorably there.

How did it get that way? Unlike the Grand Canyon, where the erosive force of a great river has revealed half of the earth's geological history, the Middle West hides much of its past. But there are clues.

The Shaping of America's Heartland takes us through two billion years of geological history to the Pleistocene, which spanned a million years and during which four great continental glaciers advanced and retreated with a profound effect on the Middle West as we see it today. During the last two or three centuries, European man, proliferating and spreading westward, modified conditions further. With his agriculture, his cities, and his roads, he has disguised but has not obliterated the evidence of natural forces at work in the shaping of the land.

Betty Flanders Thomson, Professor Emeritus of Botany at Connecticut College, where she taught for more than thirty years, is best known to many of us through her book *The Changing Face of New England* (Macmillan, 1958) and therefore we may think of her as New England born and bred. Actually, like so many other academic people living and working in New England, she was born in the Middle West. Thus, in a sense, this latest book goes back to her roots in Ohio; it does for the flat Middle West what her earlier book did for the hilly Northeast.

Miss Thomson has always felt irritated by people who found the landscape of the midcontinent uninteresting; they obviously lacked the background to interpret what they saw. There really was no book where they could find out why the prairies, the upper Mississippi River basin, and the Great Lakes region looked the way they did and what things meant. This labor of love should open their eyes.

ROGER TORY PETERSON

Foreword

THIS IS a book about the landscape of a region: what it is, how it came to be, and what it means. One of the motives for undertaking such a work was dismay at the blindness of people who think there is nothing worth seeing between Pittsburgh and Denver; for I agree passionately with Bernard DeVoto that there is no such thing as an uninteresting landscape. This is not to say that all is picturesque and charming on the till plains of Illinois or along the Red River of the North. But there is much that is intensely interesting to see in a land strewn with mementoes of a continental ice sheet, clothed once with forest and grassland of great richness, carved into mile-square sections for settlement, and now farmed on a huge and productive scale.

A second motive for the book is to make accessible to other people some of the extremely interesting things about the natural world that are known to practicing scientists. Much of the fascination lies not just in what we know, but even more in how we know it; and I have tried to convey as much of this as possible. The book is a sort of translation, for much of the information in it comes from the technical scientific literature in the fields of ecology, botany, and geology.

One impression that emerges strongly from reading widely in such literature is the love of natural scientists for the objects of their studies. Along with the graphs and tables and the statistical compilations, there are many passages of description and interpretation where this shows clearly through. The books and papers of such people as John Curtis writing on the vegetation of Wisconsin, John Weaver on the grasslands, Lucy Braun on the deciduous forest, and Paul Errington on the prairie marshes are good examples. There are many more.

I grew up in northern Ohio but moved away before I had learned much about natural history. I found that things once seen and long remembered take on a wholly new perspective when viewed again in the light of many years spent as a natural scientist living in a very different landscape. Ferreting out the information for this book can only be described as a revelation.

It will be a great satisfaction to me if my reader finds that this story helps him to see and understand more and hence find greater enjoyment in the landscape of the Middle West.

B.F.T.

Quaker Hill, Connecticut

Contents

Illustrations

Maps and Drawings

Photographs

The Shaping of America's Heartland

1. What Is the Middle West?

JUST WHAT part of this country *is* the Middle West? Probably everyone would agree that it lies somewhere west of The East, north of The South, and east of The West. Free association with the name might bring forth such words as "farmers" and "corn" and "Chicago," and perhaps "flatness" and "heat waves" and "blizzards." But on the exact geographical limits of the region there might well be some argument.

For purposes of this book the Middle West is taken as essentially the Corn Belt and the Great Lakes States. This is a large area bounded approximately by the Ohio River, the lower Missouri, the hundredth meridian of longitude, and the northern side of the Great Lakes. These are somewhat arbitrary boundaries, but they coincide quite well with natural lines of transition. A really coherent natural region would include also the neighboring parts of Canada, for the daily concerns and the natural surroundings of a farmer or fisherman in Ohio and his counterpart in southern Ontario, or the cattleman or wheat grower of North Dakota and his neighbor in Manitoba, are much alike.

This is a wide expanse of level land, not entirely flat, but with a decidedly low degree of relief. Most of it is deeply covered with glacial deposits that have developed into a deep and fertile soil. With an adequate although not entirely reliable rainfall, the country supports a highly prosperous level of farming.

Besides this level area, the Middle West must include a bit of the rocky Canadian Shield that extends south into Minnesota and Wisconsin. This is a rough country of ancient hard rocks that since the time of the ice sheet has been only thinly and patchily covered with soil. The great bulk of the shield lies northward in Canada, where it is so intractable for most aspects of modern life that it virtually cuts that country in two; only in recent years has there been more than one highway crossing it all the way from east to west. The part of the shield exposed south of the border is less formidable and soon grades off into a gentler terrain.

It is largely a matter of convenience to put the eastern limit of the Middle West at the Ohio River. It would be just as reasonable to draw the boundary along the slopes of the Appalachian Plateau. This would exclude most of eastern Ohio but include a strip eastward along the entire shore of Lake Erie and perhaps as far beyond it as Syracuse. Certainly, approaching the lake from the south in Pennsylvania or western New York one has a clear sense of coming down out of the hills into a different kind of country. The western edge of the Plateau, in central Ohio, slopes off much more gradually and is partly obscured by glacial deposits, so that one must be watching for it to perceive the change in topography. However, since the human world of the people who live there looks more westward than east, all of Ohio will be included in our Middle West.

In the lower Ohio Valley, bits and pieces north of the river that were never overridden by the ice sheets have more in common in their physical appearance with the country south of the river. But since the river marks a fairly distinct cultural as well as political boundary between Middle West and South, it will serve to delimit our region. West of the Mississippi the hilly Ozark country is a rather separate entity, and here the lower Missouri River can serve as an appropriate southern limit.

At this point tangible physical boundaries fail us. Certainly the eastern parts of Kansas, Nebraska, and the Dakotas are essentially Midwestern, but their western ends just as surely are not.

Although there is nothing on a map to put a finger on, the western edge of the Middle West, the place where the true West begins, is really quite as distinct and significant as the other boundaries we are using. The demarcation lies somewhere just east of the hundredth meridian. This geographer's convention coincides closely with a number of natural boundaries, all of them interrelated: the 2000-foot elevation line, the 20-inch rainfall line, and the edge of the tallgrass prairie. Here it is that tame-grass pastures and the farming way of life give over to wild-grass rangeland and ranching. Human habitations become farther apart, until the real rancher cannot see his neighbor's house, even in this country of clear air and unobstructed, horizon-wide views.

Northward, the Central Lowland of the continent has no physical boundary short of Hudson Bay and the Arctic Ocean. The same is true of the rough and rocky Canadian Shield. Here one can either draw a sweeping curve from Winnipeg through North Bay to Toronto to include the less rigorously northern and wild areas, or else take the easier course and use the international boundary to close the loop.

The landscape of today's Middle West is predominantly a flourishing farmland. In fact, the acreage under cultivation is still being increased by drainage and reclamation projects. Abandoned farmhouses can be found here and there, but most often it is only the house that has been abandoned, and the land that pertains to it has been incorporated into another farm to make a larger operating unit. The large tracts of abandoned farmland now growing up to woods that are such a common sight farther east are rarely found except in the cutover forest region in the north, where a brief and ill-fated attempt was made to turn the desolated wastes of the virgin forest to agricultural use. Elsewhere, land once cultivated is usually still cultivated. Most of it is so valuable for farming that it is all but impossible to find samples of woodland, except in the northern areas now devoted primarily to forest industries, or to find undisturbed prairie anywhere at all. Even the farm woodlot is a very poor specimen of woods; and the traveler looking for a spot of shade wherein to eat a roadside lunch may go a long way before he finds any accessible trees except those on a farmhouse lawn.

Before it was opened to settlement, nearly all of the Middle West was surveyed and marked out into mile-square sections. The standard farm was at first one section, 640 acres, or sometimes a half or quarter section. The pattern still shows in the network of local roads, a mile apart and running straight north-and-south or east-and-west, although it is irregularly overlaid with the successors of a few older roads of historical origin that connected important places; and of course today's superhighways ignore both the local road pattern and all but major topographical obstructions.

In recent years Midwestern farms, like farms everywhere, have been growing ever larger and more extensively mechanized. One of the regional sights is the array of mechanical monsters displayed by the dealer in farm machinery who does business on the edge of every town in the farming country: plows, cultivators of every conceivable kind, spreaders, spray rigs, sowers, hay bailers, hay driers, combine harvesters, corn pickers, corn shellers, silage choppers, and so on through all the business of farming.

Into the predominantly rural landscape cities have been expanding explosively for a hundred years. Although a number of cities originated from settlements that grew up around old trading or military posts, even as late as 1850 there was hardly a town in existence that covered an area greater than its present downtown business district. Consequently the only century-old buildings left are widely scattered individuals that managed to survive early "improvements" or else were missed altogether by the growing towns. There simply are no picturesque old neighborhoods in Midwestern cities such as one finds in Boston or Philadelphia.

My own native Cleveland illustrates the tempo of urban expansion. Although it was laid out in 1796 by Moses Cleaveland, it had little more than a thousand inhabitants by 1830. The houses were clustered around a village green that survives as the Public Square. Then came the canals and the railroads and an influx of population. When my grandmother's family came from Germany just after the Civil War, they settled in an oldish community known as "The Grove" on the western edge of the Cuyahoga Valley. My mother always remembered how she loved the small-paned windows in the little white house where she lived as a child, until in the 1890's her father built a fine new house, complete with golden oak woodwork and no old-fashioned white paint, two miles farther out on the edge of town among apple orchards. That house of my grandfather's as I first remember it was in a pleasant tree-shaded "older" neighborhood, and we lived five miles out in the new suburb of Lakewood. Now the old Grove has long since disappeared under a viaduct near the steel mills, my grandfather's house is practically downtown, and the westward-expanding edge of the built-up metropolitan area blends with the eastward-advancing edge of the suburbs of Lorain.

But the burden of this book is not cities but the land itself and the vegetation and the creatures that live on it, the things that one sees as he travels about over the landscape, and what they mean and how they came to be.

Not long ago a sabbatical leave gave me the opportunity to spend a month during the springtime making a grand circuit tour of the Middle West. For anyone interested in seeing the land and its people going about their business, and at a season when the world is fresh and men are rejoicing in the passing of winter, this is an experience to be cherished.

Crossing into the region at Wheeling, West Virginia, I went west and south to Marietta, then across the southern parts of Ohio, Indiana, and Illinois to the southernmost point of the entire Middle West at Cairo. It was early May, and all this hilly country was abloom with lilacs and apple blossoms. In the woods the ground beneath blossoming dogwood and redbud was spread with quantities of white trilliums, blue phlox, and mayapple, accented with wild geranium, bluets, and foamflower. On a clay bank in southern Indiana I made my first acquaintance with that enchanting midget, the crested iris.

In the lush bottomlands of the lower Wabash it was already early summer; the spring flowers were gone and the leaves were full grown, although still soft and pale, on the enormous maples, sycamores, oaks, and tuliptrees of Beall Woods State Park. The forest was full of birds; and the thick, adhesive mud underfoot harmonized well with the luxuriance of the mosquitos.

Northward across Illinois, the oak leaves diminished in size and the trees were instead in full catkin, and the spring flowers were again in bloom. Here were more wide expanses of rich, black fields, still very wet and some of them not yet plowable.

Still north and west into Iowa, the bright pink of wild crab blossoms appeared, and then the white of wild plum in the thickets. The steeply hilly country of this unglaciated corner would have been scenic, with some lovely long views, if it had not been obscured by heavy spring rain and mist. Up on the glacial drift sheets the real prairie country began, first strewn with old groves of oak, then with the trees confined to the sides and bottoms of valleys.

One enchanted hour I spent on a sunny Monday morning in Helmer Myre State Park, near Albert Lea in southern Minnesota. This is a maple-basswood forest that stands on the irregular surface of a moraine. It was almost as the primeval wilderness must have been — no sounds of man or his doings, only the singing of countless birds, the drumming of woodpeckers, and the vociferous calls of assorted frogs. A small, sandy road ran under trees in a mist of tiny green leaves, and here were the legendary carpets of spring flowers, expanses such as I had never yet seen of rue anemones, blue phlox, spring beauty, Dutchman's breeches, and violets both yellow and blue. A park ranger appeared briefly and went on elsewhere about his business. For all the human solitude, it was a bustlingly busy place in the natural world in that season of birth and growth.

Out on the cultivated farmlands, the first dry day after a protracted period of rain had brought all the farmers out with their enormous cultivators and planters. Here again the land was rich and black and utterly stoneless. Along the way were broad, fragrant mounds of wild plum in flower, and the trees were now predominantly bur oaks. A gloriously sweet song turned out to come from the western meadowlark, a bird of totally familiar appearance producing a totally unfamiliar sound. There must have been *zillions* of red-winged blackbirds along the way, all of them constantly in full cry with their throatily piping song. Almost as many, it seemed, were the cooing mourning doves.

On into South Dakota on the high moraines of the Coteau des Prairies, the season was less advanced (mid-May by now). In the countless ponds and sloughs, the cattails and large wild grasses and sedges still showed no green shoots penetrating last year's brown remnants. But every pond and every water-filled ditch had its pair or its small flock of waterfowl. In the bigger ponds stood the hummocks of muskrat houses. On the drier places the gophers and thirteen-lined ground squirrels scurried — there is no other word for it — busily about.

In the southeastern corner of North Dakota occurred another peaceful morning idyll at Fort Abercrombie, now a state park with a small historical museum. The fort was established in 1857 to protect the extensive trade between Canada, via the Red River Valley, and Fort Snelling, now St. Paul. The place is pleasantly tree shaded, but photos taken in the early days show a short double row of frame buildings standing naked on the open prairie. The feeling of history in the reconstructed fort was somewhat jarred for a visitor of a certain age by the display among historical

USDA Photo

Most of the Middle West is intensively farmed. Few places are as utterly flat as the floor of Glacial Lake Agassiz seen here. But the rectangular grid of the original land survey and the farmsteads scattered widely over the landscape, although close enough to be within sight of neighbors, are characteristic of the region. This view shows evidence of the many wet spots typical of primeval tallgrass prairie. Some places are still wet enough to affect the pattern of spring plowing, others show only as lighter or darker areas in the now drained fields. Thick windbreaks shelter both man and land in this wide open country.

relics of some World War II ration stamps. And the calm and quiet of the morning, heavy with a sense of the past, was abruptly shattered by the arrival of first one, then a second, and finally a total of five busloads of schoolchildren, all full of youthful and noisy ebullience.

Then back into Minnesota, skirting the edge of the wooded country. Here every marshy spot was bright with stems of red osier dogwood and lit with the shine of marsh marigolds. With the trees just budding forth, the countryside was green and fresh after spring rains. It was also warm — shirtsleeve weather with a hot southwest wind rapidly drying the fields and letting the farmers at last be about their spring plowing.

Then north and east from the fertile farmland of Glacial Lake Agassiz to what is still largely wild prairie and semi-muskeg, and always the season becoming earlier spring. At the Agassiz National Wildlife Refuge, identified on some road maps only as Mud Lake, a brief stay at a breeze-swept observation tower showed miles-wide expanses of alternating marsh vegetation and open water. Waterfowl of every description were recognizable by even a casual birdwatcher. A family of Canada geese sailed by, one parent at each end of an Indian file of six downy goslings. There were white-fronted geese, ringnecked ducks, shovelers, redheads, scaup, and pintails. There were pert little ruddy ducks, pied-billed grebes, and Franklin's gulls. Here, too, were the universal redwings; the yellowheaded blackbirds so improbable to an easterner who had never heard of such a thing; and the black terns, neat gray birds with black wings, swooping and sweeping in a fashion that, for inland birds, suggests swallows and caused no end of confusion to an outlander.

The brushy prairie along the road that follows an old Lake Agassiz beach ridge toward Lake of the Woods was still in its somber winter aspect. There a fleeting glimpse of a lone coyote gave a feeling of wildness to the scene.

Along the Rainy River it was Sunday afternoon of the first warm weekend of the year, barefoot-hot for children not long released from the restraints of winter, and a day for boating and fishing and family picnics in a landscape where the leaf buds were just starting to swell and the grass barely showed green. This bucolic scene was once the route of the canoe brigades, of the voyageurs, and the fur trade with the far northwestern wilderness.

It was also a passage through the transition zone between prairie and forest as bigger and bigger groves of aspen and birch appeared, at first in pure stands, then with a distinct understory of spruce, fir, and white cedar. At International Falls the boreal forest began, and with it rocky outcrops of the old Canadian Shield — as abrupt a transition as one could wish to see.

On the long ride home across Minnesota, northern Wisconsin, and upper Michigan, it was north-woods country, with spring progressing again through hepaticas and marsh marigolds to anemones, violets, and trilliums. After the long circuit of prairie and then evergreen forest, the deciduous woods and farmlands of southern Wisconsin and lower Michigan, now in full flood of late spring, seemed almost tropically luxuriant. The great spectacle of late May through this land seemed to be trilliums, trilliums by the acre and by the million, all the way from Duluth to Ann Arbor and beyond.

Three tangible things stand out in the memory of that trip: the cultivated earth, fertile, dark, and stoneless for hundreds and hundreds of miles, the gift of the ice sheets; the red-winged blackbirds in full plumage and song, millions of them, denizens of the endless numbers of marshy places large and small, another gift of the glaciers; and the woodland carpets of wildflowers, most especially the great white trilliums.

Less tangible but profoundly moving is a sense of the great beauty and richness of this land. The desecrations of pollution and urban blight notwithstanding, vast areas are still essentially unspoiled. The astonishing thing is that not even the many years of thriving agriculture and industry, with incessant cutting, clearing, and cultivation, have managed to eliminate the native flora and fauna, and each part of the Middle West still has its own distinctive look.

2. The Lay of the Land

ALTHOUGH THE Middle West is by no means so flat as is commonly believed, one must acknowledge that its general aspect is predominantly and conspicuously level. It was not always so, for the country has had its ups and downs; but in the long history of the region the three major events that shaped the present landscape have all had generally leveling effects.

The first of these took place in the remotest times of which we have any firm knowledge, the Archaean or Precambrian epoch. So incomprehensibly long was this period that within its span the earth was repeatedly folded into mountains, covered with water-laid sediments, intruded with molten rock, and eroded again to flatness. The complexities of those long-ago contortions and abrasions of the earth have not been completely deciphered. In the Middle West the rocks that record them appear only in northern Minnesota and Wisconsin and are largely concealed beneath deep deposits left by the later ice sheets. However, one of the last Precambrian events is revealed in the rocks that surround the western part of Lake Superior.

Along the north shore of the lake, from Canada and Isle Royale to Duluth and on inland, the land rises sharply, exposing the edges of layer upon layer of basalt, a rock composed of congealed lava. The basalt of this region has a dark gray or purplish red color that gives a rather somber cast to ledges, cliffs, and beach sands formed from it. The face of the great basalt exposure makes a dramatic setting for the city of Duluth, which rises six hundred feet from the water's edge to the upland surface only a mile or so inland.

The massive rocks that rim the lake shore here are one side of a huge trough whose bottom lies deep beneath the lake floor. Beyond the lake its other side rises to form the Copper Range, the Porcupine Mountains, and the Keweenaw Peninsula.

This trough is part of a huge mass of lava that emerged from long rifts in the earth in late Precambrian times. At that time the land was broadly level, for the mountains that had arisen earlier had been almost entirely worn away. Onto the level surface, flow upon flow of molten rock poured forth and spread far over the land. Geologists believe that such rifting and outpouring of lava may have gone on as far away as Kansas and lower Michigan.

In any case, the amount of lava was immense. Some individual flows are known to bulk as large as a hundred cubic miles, and the combined thickness of them all is estimated at twenty to thirty thousand feet — four or five miles. The landscape of those times must have looked much like the present plateau region of eastern Oregon and Washington, or like Iceland or the Deccan plain of India.

During later stirrings of the earth, the entire vast mass of lava sheets became bowed down into a great trough. This was not yet the basin where Lake Superior now lies, but an ancestral lowland that became filled with sediment as the waters running over the earth went about their eternal business of taking from the high places and putting into the low. In time the accumulated sediments became transformed into sandstones and conglomerate rocks.

At last the entire land lay quiet for so long, while running water wore away at the roughnesses, that only a scattered remnant of hills and undulations interrupted the general flatness of the primordial Middle West. Into this scene had been built the potential for making a rugged landscape, since there were differences in the hardness of the various layers touching the surface. But for the time the roughness was latent.

So ended Precambrian times, five hundred million years ago, and the first major shaping of the landscape.

Then came the inland seas of Paleozoic times. Through much of the next three hundred million years, the entire flattened center of North America was alternately flooded by the sea and drained again. Each time the land was submerged, it acquired a thick, smooth cover of sediment, and each time it stood again above the water its surface was roughened by erosion, leaving a record of the events for the edification of geologists far in the future.

Meanwhile, far to the east, repeated intervals of mountain-making gave rise in succession to the Taconic, the Acadian, and finally the Appalachian Mountains. All this disturbed the land farther west very little, however, and there the thick layers of level rock were merely flexed into broad, shallow basins that now lie centered under Illinois, lower

Cross-bedded Paleozoic sandstone exposed in a road cut beside Lake Pepin, Wisconsin. Sand particles weathering out of the bedrock show how soft these rocks are.

Michigan, and the hilly country of southeastern Ohio and nearby Pennsylvania and West Virginia.

In the wide, flat basins, deposits accumulated that have become the natural resources of today. During some parts of the long ages, when the climate was warm and wet, there were large expanses of shallow water where vegetation grew thick and lush. On the swampy bottoms the dead remains of this luxuriance accumulated in such quantities that even after they have been compacted by the weight of thousands of feet of overlying rock and have been chemically altered so that little is left except carbon, the resulting coal still lies in beds and seams many feet in thickness. Elsewhere the shells of countless millions of minute creatures settled to the bottom of the warm, shallow water, to be converted in time to great thicknesses of limestone.

At other times the climate was highly arid, and evaporation was intense. Then parts of the diminishing inland sea became cut off from the main body of water and completely dried away, leaving beds of salt and gypsum.

At the end of the Paleozoic the eastern half of North America stood above the sea, never again to be submerged except along its coastal fringe. Far to the west, seas of a later period would reach as far east as the present High Plains, leaving their mark in the form of hard sandstones that still cap mesas and escarpments. But for the past hundred million years the Middle West has remained dry land.

Through all this time the surface of the land has been constantly eroding, sometimes exceedingly slowly, sometimes spurred on by periods of gentle uplift. One of the largest uplifts raised the entire center of the old Precambrian land mass and brought about the removal of its soft Paleozoic mantle, exposing the long-buried hard rock plain. The center of this great updoming lies to the north and forms the Canadian Shield. Only a small part of it reaches as far south as Minnesota and northern Wisconsin, where it is known as the Superior Upland.

As erosion worked down into the ancient surface, differences in hardness of the various rock formations began to exert an influence on the developing landscape, and in the process many a long-buried feature has come to light. One of these is the lava-bedded trough that underlies western Lake Superior as well as the St. Croix Valley south of it. On both sides of the trough the uptilted edges of the lava beds rise as outfacing escarpments that run for many miles. The center of the trough is floored with newer, flat-lying sandstones and conglomerates. Much of these have been eroded away, but the more resistant central parts remain to form the spine of the Bayfield Peninsula and the Apostle Islands.

Away from the lava trough the exposed edges of other hard layers among the ancient folded and tilted rocks stand out as higher ridges. Such are the iron ranges, Mesabi and Vermilion in Minnesota, Gogebic and Menominee in Michigan along its border with Wisconsin. Such also are the low divides between the long, narrow lakes of the Gunflint district in far northeastern Minnesota.

The Huron Mountains in northern Michigan had a different origin. They consist of a large mass of granite that formed deep within the earth, below the soft rock mantle that has since been removed from it. This was once a smoothly rounded boss, but in the ages since it was uncovered, rivers have incised it with sharp but shallow valleys.

West of Lake Superior the underlying rock surface is both hard and fairly level. Most of it is hidden by deep glacial deposits, but it is believed to be a remnant of the old Precambrian flatland. The country is a maze of swamps and lakes, drained, more or less, by the St. Louis River. Northeastward, in the "Arrowhead" country of Minnesota and the valley of Rainy River, the glacial cover thins out and much rough and irregular bedrock is exposed. This is a beautiful northern wilderness of lakes and evergreen forests. A part of it has been set aside as the Boundary Waters Canoe Area.

Going in any direction from the Superior Upland, east,

south, or west, one comes out onto the level surface of younger Paleozoic rocks that lap up on the edges of the ancient shield. Where south-flowing rivers cross the margin, they descend abruptly over falls or rapids from the hard upland surface into channels cut deep into the softer, younger sediments. Such are the falls and gorges of the St. Croix and Chippewa Rivers and the Wisconsin River Rapids.

In contrast to the dark color and contorted form of most of the Precambrian rocks, those of the Paleozoic are light in color and soft in texture, and so little are they disturbed from the horizontal position in which they were laid down that only large scale observations and fine measurements show any slope at all. To a casual observer they appear perfectly level and flat.

The oldest of these rocks, the Cambrian, are widely exposed around the eastern and southern fringes of the Superior Upland. The Pictured Rocks and other bluffs along the Michigan shore of Lake Superior, the Dalles of the Wisconsin River, and the high cliffs along the Mississippi River trench between Wisconsin and Iowa all show the ashy-pale or softly golden color and the clearly visible bedding and cross-bedding of these old sediments. Some of the rocks are so soft they crumble into sand under a probing fingertip, and it is clear that they have suffered little from geological vicissitudes since they settled out of the ancient seas.

Although they are all relatively soft as rocks go, the Paleozoic layers differ considerably in their resistance to erosion, and such ups and downs as the landscape shows are due in large part to outcropping edges of the more durable strata. In fact, one layer of hard dolomitic limestone is largely responsible for the overall geography of the Great Lakes region. This is known as the Niagaran formation because it appears at the lip of the great falls and, indeed, is responsible for their existence. It has the shape of a broadly rounded saucer whose center lies at a depth of eight thousand feet under the middle of lower Michigan. This we

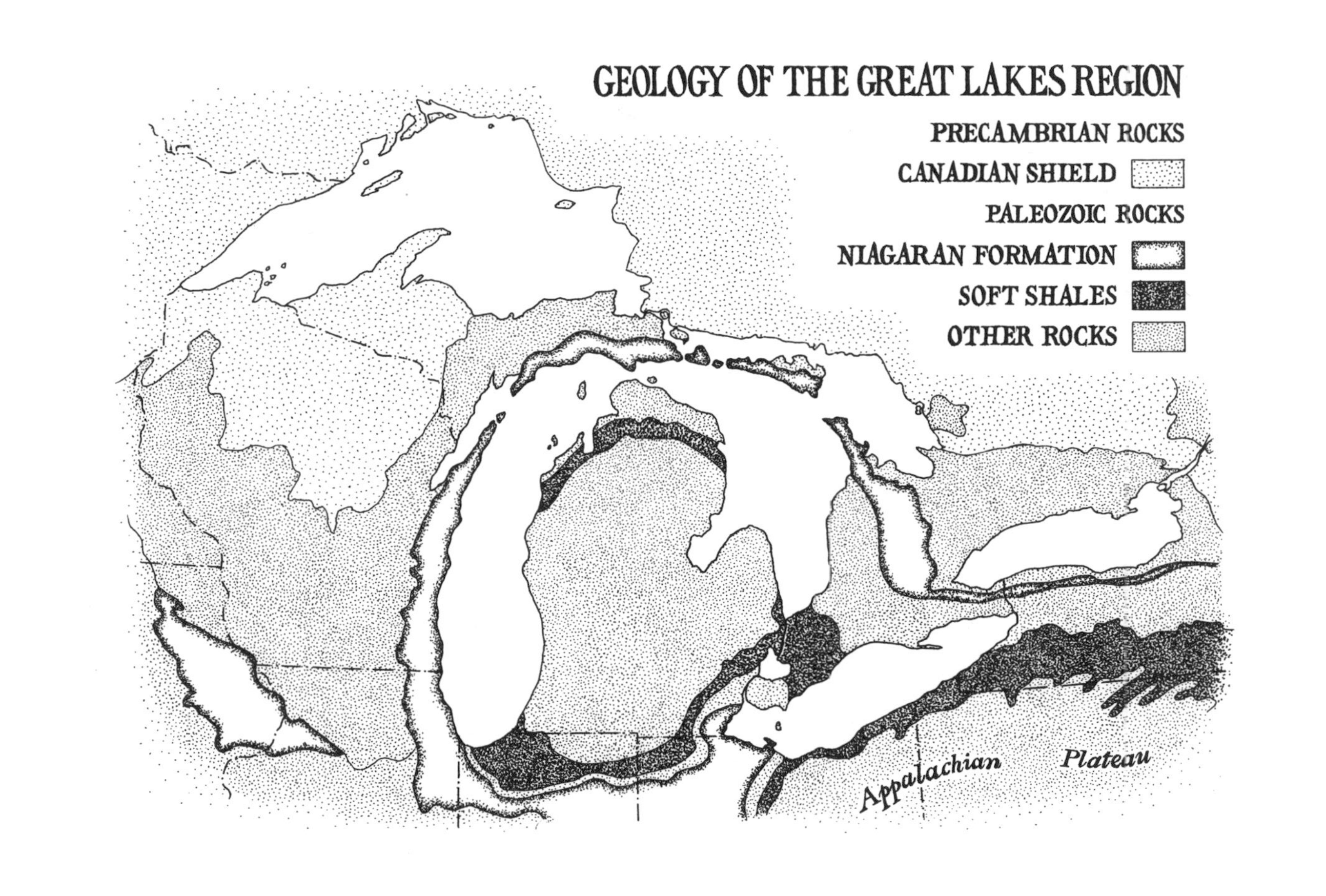
GEOLOGY OF THE GREAT LAKES REGION
PRECAMBRIAN ROCKS
CANADIAN SHIELD
PALEOZOIC ROCKS
NIAGARAN FORMATION
SOFT SHALES
OTHER ROCKS
Appalachian Plateau

know from the drilling records of a large number of wells and mines.

Within the Niagaran limestone lies a series of progressively younger rock strata whose edges have been smoothed off rather like the tops of a set of nested kitchen bowls. Outside and below the Niagaran lies another series of progressively older strata. Among all these layers, the hard Niagaran protrudes as a ridge or escarpment wherever it touches the surface.

On a map, the Niagaran escarpment appears as a great loop around the lower Great Lakes. It forms the southern edge of the basin of Lake Ontario, extending across New York State and into Canada at Niagara Falls. West of the falls it veers sharply northward to make the peninsula and islands that separate Georgian Bay from the main body of Lake Huron. It curves around the northern side of the Straits of Mackinac into the Upper Peninsula of Michigan, then down the western side of Lake Michigan, making the peninsula of Door County in Wisconsin. From Chicago it swings southeast and then east across Indiana and into the northwest corner of Ohio. There it dips into the earth near the city of Sandusky after leaving some last low, rocky ridges protruding through the glacial drift. One of these forms the spine of Catawba Island and the Bass Islands in western Lake Erie and provides a platform for the old lighthouse at Marblehead.

Just within the Niagaran limestone lies a zone of particularly soft shales. These have been deeply scoured to form the basins of the middle three Great Lakes — Erie, Huron, and Michigan. Just outside the Niagaran another soft-rock zone forms the beds of Lake Ontario on the east, then Georgian Bay and the North Channel leading to Lake Superior, and on the west Green Bay, Lake Winnebago, and the upper parts of the Rock and Illinois River valleys.

Westward in Wisconsin the Paleozoic rocks thin out and disappear against the Superior Upland. Near the thin edge of the oldest and lowest of them, protruding like a half concealed fossil, is the Baraboo Range. This double ridge of old hard rock once stood as an isolated remnant on the Precambrian plain, until the rising seas engulfed it and buried it in sediment. During the long span of uplift and erosion since the sea has disappeared, most of the younger rocks have been stripped away, and the old ridge once more stands forth almost as it must have done those hundreds of millions of years ago.

Outward from both the Michigan Basin and the Superior Upland the Paleozoic rocks arch gently upward and then down and disappear. On the east, in Ohio, they merge into the edge of the Appalachian Plateau. Southward they sag beneath the younger sediments of the Gulf coastal plain. Toward the west they are overlain by the sharp edge of the still younger rocks of the High Plains. On the southwest rises the Ozark Plateau, another Precambrian mass, and onto the edge of this the Paleozoic rocks taper off just as they do on the Canadian Shield.

Such, then, is the bony structure, the bedrock shape of the Middle West. Into the gentle lifts and sags of its surface, rivers and streams long ago carved a pattern of valleys that would carry the waters off to the Mississippi, the St. Lawrence, and some ancestor of Hudson Bay. But all the surface details and even some of the larger features were to be drastically altered by a third event, the coming of the continental ice sheets.

In geological terms it was not long ago, only a million years or so, that the ice first came down from the north. Four times it came and four times it has melted away, the last time just the flick of an eyelid ago, a mere matter of ten or twelve thousand years.

The soft rocks of the midcontinent region were easy prey to the grinding action of the massive, moving ice. In the far north, where the bulk of the ice came from and where it was thickest, large areas of the Canadian Shield were totally denuded of soil and probably lost also many feet of bedrock. Areas that still had their soft Paleozoic covering were even more deeply eroded.

What was taken from one place was eventually set down

somewhere farther south in the form of glacial drift.* Most of the material gathered up in the north ultimately came to ground in our Middle West, where the thinning outer edge of the ice lost its erosive power and began instead to lay down its burden. There over great expanses of the lower, flatter country the ice spread layer upon layer of drift to a thickness that often reached several hundred feet.

A thin coating of drift may have little effect on the landscape, merely softening contours by rounding off hilltops and filling the deepest valley bottoms without completely obscuring the original form of the land. But the deposits over much of the country from Ohio to Iowa are so deep that they make an entirely new surface that bears no relation to the land buried below.

Now, in geological terms the million years since the first advent of the ice is a brief interval, not long enough to make any major change in the surface of a stable land under ordinary circumstances. So it seems safe to assume that the country just beyond the farthest reach of the ice still resembles the hidden preglacial landscape. In the Middle West, this means southeastern Ohio, a part of southern Indiana, the extreme southern tip of Illinois, and the strange island-like area in southwestern Wisconsin and the northwestern corner of Illinois that none of the various ice sheets ever quite covered.

In the north, the unglaciated part of Wisconsin and Illinois is a broadly rolling upland plain that has been deeply cut by streams into a fine textured network of valleys and ravines. In some places cliffs, mesas, and sharp pinnacle-like rocks stand boldly out from the surrounding terrain in forms that could never survive the attack of a full-fledged ice sheet. In general, enough is left of the old upland surface to provide almost continuous passage along the ridges between valleys, the "interfluves" of geologists, for a system

* *Drift* is a useful geological term meaning anything that has been transported by ice, whatever else may have happened to it.

of main roads. Lesser roads tend to follow the lesser ridges between tributary streams, dipping down here and there to cross the head of a smaller valley, while small side roads run steeply up and down the length of the smallest side valleys. The entire region is strikingly different from the more level drift-covered country around it, as even a casual traveler will notice.

South of the ice margin, near the Ohio River, the country is more extensively eroded, valleys are deeper, and little or nothing remains of the original upland surface. Most of the interfluves no longer form flat-topped ridges but have been carved into broken and irregular hills.

What lies under the drift that covers the rest of the region has been reconstructed in surprisingly fine detail. The chief source of information for this is the drilling records of thousands of deep wells. The preglacial contour map that emerges shows a valley-carved land with a drainage pattern widely different from the present one.

One of the most spectacular changes was the obliteration of a great river that once swept in a wide arc from the mountains of North Carolina far across northern Indiana and finally entered the head of what was then a long arm of the Gulf of Mexico, in the vicinity of the present St. Louis. The ancient river has been named the Teays, for a little town standing by it in West Virginia.

The headwaters part of the Teays, from a point near Blowing Rock to the present Kanawha, remains as the present-day New River. A few other bits of its valley carry segments of modern rivers — the Ohio between Huntington, West Virginia, and Wheelersburg, Ohio, the Scioto below Chillicothe, and the Illinois below Beardstown. All the rest of its huge valley has been completely filled with a combination of glacial drift and lake-bottom sediment. In the north, where the entire Teays drainage system was overridden by the ice sheets, it is ice-laid drift that obliterated the valley. The rest disappeared under ice-margin lakes.

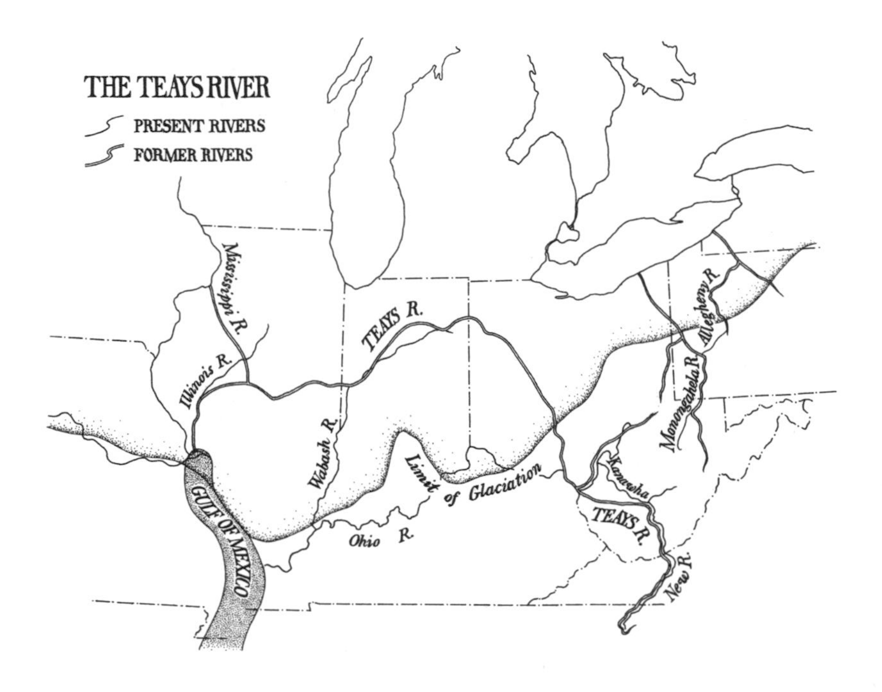

When the glacier bore down into the old valleys, rivers flowing north through them were blocked by the advancing face of the ice. The water that continued to come from upstream then backed up to higher and higher levels, spreading broadly until finally it overflowed through some gap in the surrounding hills. In this way many ice-margin lakes, both large and small, came into existence. Although they were in the long run temporary, the lakes lasted long enough to acquire fillings of sediment that in some places are two hundred feet deep. The smooth surfaces of such old lake bottoms, flat and fertile, are a conspicuous feature of the landscape south of the glacial border, not only in the Teays Valley of southern Ohio, but west to the Mississippi and beyond.

In the former domain of the Teays the master river is now the Ohio. This is a composite of several rather unimpressive preglacial rivers that became connected by completely new postglacial streams. The lower part consists of a river that probably headed originally somewhere in southwestern Ohio and followed essentially its present course to the Mississippi. The headwaters streams, the Allegheny and Monongahela, seem to have been separate rivers that flowed in what is now their upstream direction and ultimately into some ancestral St. Lawrence Valley. The middle reaches were put together from the deepest parts of overflow streams that developed on the ice-margin lake floors as the country gradually drained.

Other rivers of the Middle West were remodeled by the glacier. One of these was the Mississippi. This was less drastically affected, however, being merely diverted from part of its old course. Its upper reaches as far downstream as Rock Island, Illinois, were largely the same in preglacial times as they are now; but below this the old Mississippi flowed southeastward to join the Teays near the present town of Lincoln. Early in the Ice Age a glacial lobe moving down from the trough of Lake Michigan blocked that part of the channel and the river shifted farther west.

Another great river whose course was drastically altered by the ice sheets is the Missouri. In preglacial times the lower Missouri joined the Mississippi even as it does today; but its upper basin above about the center of South Dakota appears to have drained eastward and then north to the Arctic by an ancestral Red River. Moreover, it is clear that in the Dakotas the main river lay farther east than it does now. A look at a modern map suggests that there is something odd about the drainage of that region, since the Missouri receives many tributary streams from the west but only a few from the east, and those are rather inconsequential. On the ground, the channels of those east-flowing rivers can be traced right across the Missouri Valley and beyond. East of the main valley they are clogged in places with glacial deposits, but not beyond recognition. Such streams as they carry are not very large and now flow west into the Missouri.

It appears that here again the main valley was overridden

head-on by the ice and became covered with glacial drift. The drift here is much thinner than it is over the Teays, however, and does not totally obscure the old valleys. The main river follows more or less along the outer edge of the glacial deposits and cuts at right angles across its old tributaries. The flow in this section of the river has also been reversed, although it is not clear just where the old divide between Arctic and Gulf drainages actually lay.

The ice in a sense also remodeled the Great Lakes that so dominate their corner of the Middle West. It is quite certain that these did not exist as lakes before the Ice Age. But the soft rocks that were to be gouged out by the ice to form water-holding basins had no doubt been found by runoff waters and carved into a system of valleys. Details of these valleys are largely conjectural, although some have been identified, such as a large one running through the Straits of Mackinac. It seems certain that the whole area drained to the northeast very much as it does today.

Around the fringes of this glaciated country there are many places where an amateur can enjoy a little glacial border hunting. A north-south traverse around Canton or Lisbon, Ohio, or north of Bloomington, Indiana, will cross the outer ice margin. In Wisconsin one must go east and west, as at Baraboo or between Madison and Mount Horeb.

What does one look for? What are the signs that reveal just where the ice has and has not been? On a large scale, there is the difference in contours of the landscape. In the lower Middle West the most striking effect is the filling of valleys and general leveling up that has already been mentioned. Farther north this is compounded by the conspicuously irregular jumble of ridges, hummocks, and hollows of massive end moraines; all this will be part of another chapter.

On a small, local scale probably the best clue is the nature of the soil and the stones embedded in it or lying on top of it. Any place in this region where the surface is strewn with rocks bigger than pebbles has almost certainly been glaciated. Glacier-borne rocks have a rather worn look from being ground against bedrock or against other moving rocks during their ride in the ice. Often their sides have been worn flat, though the corners are rounded. Rocks derived in place directly from the bedrock beneath them usually have sharper edges, unless they have been smoothed by tumbling with other rocks in the bed of a stream or on a wave-washed shore. Then they are more regularly rounded and lack the flattened facets of ice-transported stones.

A fresh vertical cut into the soil reveals other differences. Where the surface soil has been formed from the solid bedrock beneath it by ordinary processes of weathering and breakdown, the whole sequence of change from unaltered bedrock to mature soil can be seen, provided the cut is deep enough or the soil shallow enough. Starting from solid mother rock and working upward, scattered cracks and cleavages begin to appear, far apart at first, then carving the rock into smaller and smaller blocks. The smaller blocks in turn become "rotten" and appear to be disintegrating. Sometimes changes in color attest to the progress of chemical changes that go on under the influence of air and water penetrating from above. Nearer the surface the rock becomes reduced to single grains the size of sand, silt, or even clay. In such a place, if the soil is very shallow, small stones may be mixed into the surface of a cultivated field where the plow nicks into the rotting rock below.

The upper part of the soil is progressively more influenced by plants growing on it and insects and other animals burrowing into it. The uppermost layers are darkened by the decayed remains of many generations of dead roots and other parts of plants that gradually filter down among the rock particles. The topmost inch or so may contain almost as much humusy organic matter as it does rock fragments.

In a recently glaciated land the soil shows none of this orderly change. To be sure, the growth and death of plants has the same darkening effect on the surface layers, and the

Deep loess near Broken Bow, Nebraska. The vertical sides of the road cut hold their shape with neither support nor vegetational cover.

strictly chemical effects of weathering may produce color changes below the reach of roots; but the gradual transition from the underlying solid rock is entirely lacking. Instead, where the soil contains any rocks at all, they form a chaotic jumble of pieces of all sizes, with big and little stones mixed into the finer stuff at random, like the fruit and nuts in a fruitcake. Glacial drift derived from soft rocks may be ground so fine that it contains no stones at all, right down through the great depth to bedrock.

Much of the material laid down directly by the ice has subsequently been rearranged by water or wind. Water-laid deposits, whatever their ultimate origin, are sorted by size, for particles settle out in layers according to the speed of the water that carried them. Only the fastest water is capable of moving large cobbles or boulders, while fine clay settles only in the very quietest water.

Large areas of the Middle West are blanketed with the wind-laid material known as loess, most of it derived from glacial drift. Loess is a homogeneous kind of fine-textured silt. It is quite firmly compacted, like hard drifts of wind-driven snow, and holds together in a blocky form that is remarkably resistant to ordinary erosion. Where roads are built over a deep layer of loess, cuts made with vertical walls or carved into sharp-angled steps remain intact for years, even when the surface is left entirely bare.

All things considered, the continental glaciers had a profound effect on the Midwestern landscape. So widespread, so diverse, and so clearly visible are the effects of their passing that those mighty sheets of moving ice deserve a chapter for themselves.

GEOLOGICAL TIME SCALE

Years since beginning		
12,000	Postglacial	
50,000	Wisconsinan	Pleistocene
300,000	Illinoian	
700,000	Kansan	
1,000,000	Nebraskan	
65,000,000	Tertiary	
500,000,000	Mesozoic	
600,000,000	Paleozoic	
>2 billion	Precambrian	

3. The Wake of the Ice Sheet

WE REALLY do not know just what triggered the growth of the continental glaciers of Pleistocene times. Whatever the cause, it was on the high land that stretches from Newfoundland to Baffin Island that the snow first began to accumulate. Year after year it grew deeper and deeper, gradually compacting under its own weight until it recrystallized into ice. In time, the lower part of the growing mass softened under the increasing pressure upon it, and, like a thickly viscous liquid, began to flow.

Like any fluid, moving ice takes the path of least resistance, following whatever lowlands and valleys it encounters. East of the Appalachians, where the land is relatively low, the ice moved broadly down over New England, reaching as far as Long Island and northern New Jersey. The higher land of western New York and Pennsylvania presented more of an obstacle, and there the ice sheet stalled along the northern edge of the Appalachian Plateau. Beyond this the wide, level expanses of the Middle West offered no hindrance, and there the ice spread far to the south, its movement further eased by the slope of the land toward the Gulf of Mexico.

Two other tracts of land were high enough to deflect the oncoming ice. One of these was the old Precambrian upland of northern Wisconsin that stands between the lowlands where Lake Superior and Lake Michigan now lie. As the ice moved down the lowland troughs, it seems to have divided around the higher land and, moving on, never quite closed in again over the southwestern part of Wisconsin and the small corner of Illinois next to it. The region may never have been an actual island in the ice; but the various glaciers moved past it on one side or the other, and all the land around it was glaciated at some time in the past. It is not certain that the area was never covered with ice at all, since there is some question of a glacial origin for certain rocks and rubble there; but the effect was slight and any traces that remain are not readily apparent. The region is commonly known as the Driftless Area.

Another tract of high land forms the Coteau des Prairies, or Crest of the Prairies. This is a long, high ridge that roughly parallels the eastern edge of South Dakota. The earlier ice sheets closed over the top of it and moved on, but later ones never rose so high, and their outer moraines are piled along its flanks, making some exhilaratingly high, wide, and handsome country.

Westward, where the land rises to the High Plains and the climate is drier, the ice margin swung far north, until finally it met the glaciers coming down from ice caps on the high mountains.

There is every good reason to believe that the ice spread over the Middle West four different times. Three times it has melted completely away, and as of now the fourth major melting has done away with virtually all but the Greenland ice sheet in the northern hemisphere.

The first two of the glaciers spread the farthest and reached just about the same outer limits. The earliest, or Nebraskan, is known only from the drift * it left. This is entirely covered by later deposits, but it can be recognized

* "Drift" is anything that has been transported by ice. The drift set down directly by a glacier in the place where it now lies is known as "till." This is an old Scottish word that refers to a stiff bouldery clay; it has been taken over by geologists as a technical term. Drift that is later transported by water or wind remains drift, but it is not till.

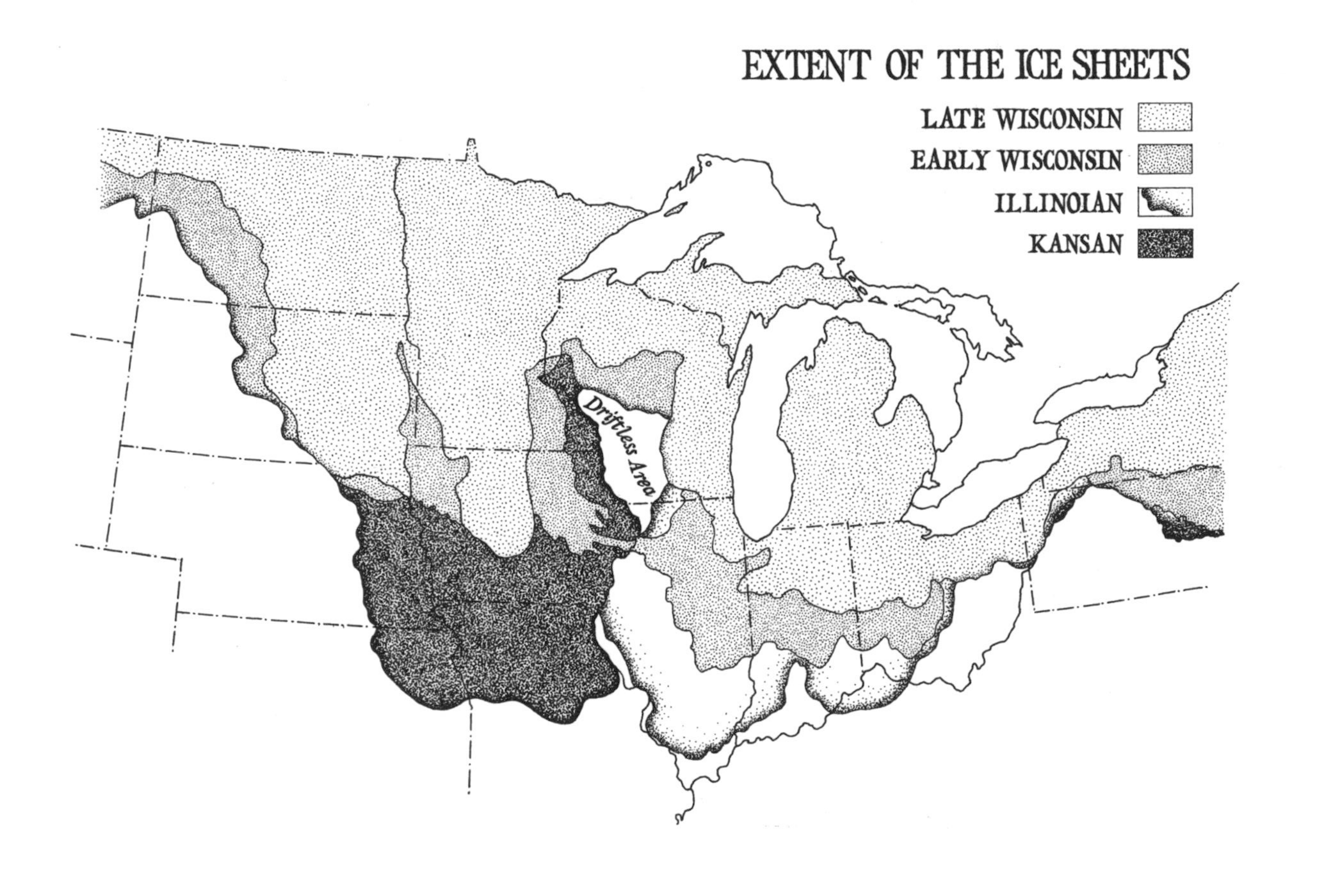
EXTENT OF THE ICE SHEETS
LATE WISCONSIN
EARLY WISCONSIN
ILLINOIAN
KANSAN
Driftless Area

in the sides of stream valleys, well cores, and other excavations into the earth.

Drift of the second ice sheet, the Kansan, forms the land surface over a wide area in northeastern Kansas and adjacent parts of Nebraska, Missouri, and Iowa as well as a strip along the entire western side of the Driftless Area. It has been exposed to the winds and the rains so long that it has lost much of its original postglacial character. Moraines and other glacier-built features have completely disappeared, and the old drift surface has been carved with stream valleys as much as four hundred feet deep. These drain the country quite effectively, and there are none of the swamps, lakes, and waterlogged places so typical of newer glacial deposits. Most of the area is covered with windblown loess from later times.

The Illinoian drift lies exposed over about half of its namesake state as well as in southern Indiana and a rather narrow strip across Ohio. About half of the original drift surface has been cut away to form a network of steep young valleys that are now about fifty to a hundred feet deep. Between the valleys the land is remarkably flat, for most of the old ice-formed irregularities have eroded away.

It is the last of the ice sheets, the Wisconsin (more properly but less pronounceably the Wisconsinan), whose work is most vividly clear. The scant ten thousand years since the last of it melted away from the upper Middle West has brought little change to the land, and all the ice-wrought details are sharp and fresh. Even small events remain clearly recorded, so much so that the history of this glaciation is known in a rather complex degree of detail.

Of all the marks that an ice sheet leaves upon a land, some of the smallest, but in a way the most vivid, are the parallel scratches or striae that were cut into bedrock by other rocks carried along in the underside of the ice. As it moves over the land, a glacier picks up a great load of debris of all kinds and sizes. This becomes pushed up into the pressure-softened underside of the ice. Often the lower

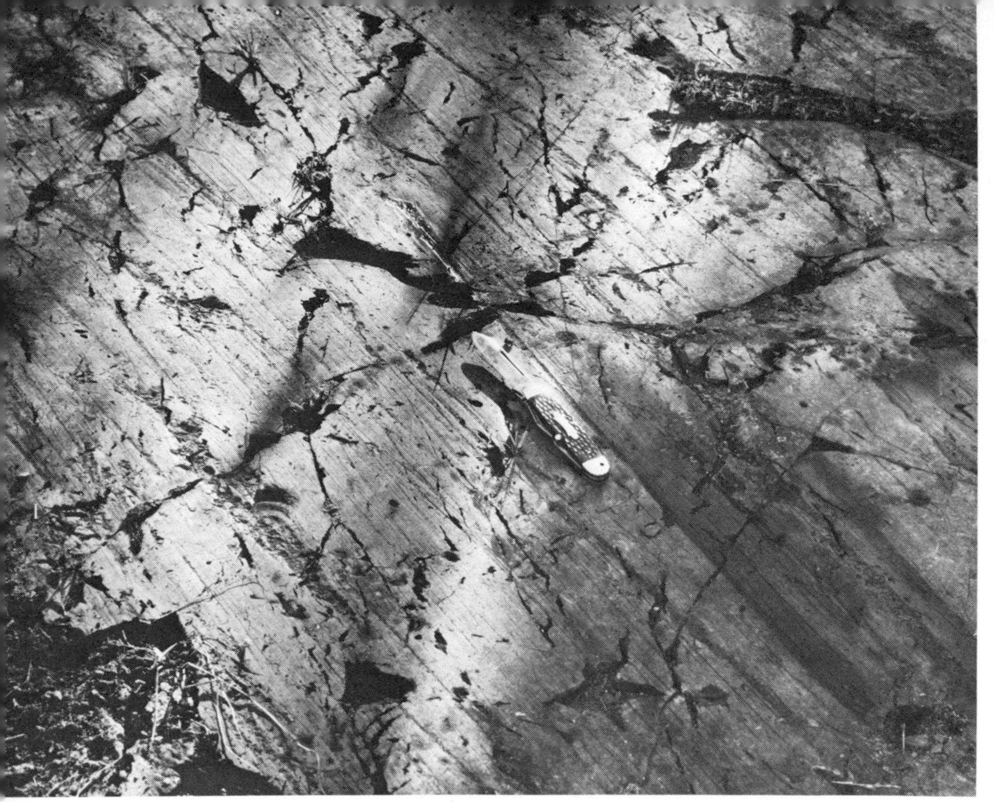

U.S. Forest Service

Precambrian rock of the Canadian Shield near Ely, Minnesota. The exposed surface of this hard rock was polished and then scratched by the ice sheets moving over it. Sharp details of the markings were preserved for thousands of years by a covering of silt.

part of a glacier contains more rock than ice. When such a rock-clad sole grinds over the earth, heavily weighted by the enormous mass riding on it, even the hardest surfaces may be scoured and scraped.

On smooth, hard bedrock, glacial scratches are sometimes so delicate that they show up only after the surface has been swept clean and wet down with a bucket of water. Soft rocks, on the other hand, may be fluted with deep, gutter-like grooves. On the limestone of Kelley's Island in the western end of Lake Erie the effect is so striking that the place has been preserved in Glacial Grooves State Park.

Perhaps the greatest of the glacier's erosive work that is still visible in the Middle West was the gouging out of the basins of the Great Lakes. These almost certainly existed in preglacial times in the form of river valleys, but the masses of ice moving lengthwise up those soft-rock channels must have enlarged them enormously.

The most massive erosion wrought by the ice is no longer visible as such at all, for the many feet of bedrock scraped from the land were ground as though in a mill to form what was later set down as drift. Rocks gathered up into moving ice become abraded and pounded and crushed, not so much by the ice as by all the other rocks around them. The hard granite and gneiss of the north and east are relatively resistant to this treatment; but the softer sandstone, limestone, and shale of the midcontinent were both easily eroded and readily pulverized.

In this part of the world the most spectacular effect of the long glaciation was not erosion but rather the subsequent spreading of layer upon layer of drift over thousands of square miles of the land. Geologists estimate that over the Great Lakes region as a whole the drift averages some forty feet deep. Over Iowa the average approaches two hundred feet; and a number of places are known where it is much deeper than that. One buried valley in Ohio lies under 762 feet of drift, and recently a new record of 1189 feet was set at a place in Michigan just south of Cadillac.

It is this truly prodigious quantity of ice-deposited rock debris that forms the level surface of the present Middle West. Here, instead of coming down from the hills in the traditional way, the traveler over the land climbs up out of the hills onto the level top of the wide glacial flatlands.

Contrary to common belief, evidence shows that the great bulk of glacial drift actually originated not so very far from where it now lies, although material laid down by one glacier may later be picked up and carried farther by another. The dark red rocks so common in the Lake Superior country produce a characteristic red drift that can be readily recognized and is very useful in the study of such matters.

A few cases are known, however, of really far-traveled things. Chunks of pure metallic copper that could only have come from the Keweenaw region of far northern Michigan have been found as far afield as Iowa, southern Illinois, and eastern Ohio.

More mysterious and stimulating to the imagination are the eleven pebble-sized, high quality diamonds that have been found in glacial drift at widely scattered spots from Wisconsin to southern Ohio. The most eager study of all the signs showing where the drift came from leads only to a rather indefinite region somewhere between Lake Superior and Hudson Bay. Since most of this is wild country, neither well known nor easy of access, the wilderness still keeps its secret against the dreams of treasure hunters.

The entire mass of drift set down by a single glacier has the form of a vast and irregular blanket. Large parts of this may be smooth and level, with only faint undulations on its surface; but other parts are molded into quite elaborate shapes.

One of these is the low, streamlined hill known as a drumlin. Just how drumlins are formed is a matter still being argued by geologists, but they seem to be made by some process that goes on beneath the ice, and it seems to happen only in certain places. Single drumlins are practically unheard of. They appear rather in flocks of dozens or hundreds. One of the famous drumlin fields of the world, in southeastern Wisconsin, has about five thousand of the little hills. There are many more in the upper part of Minnesota.

Drumlins come in different sizes, but a fair sample might be a mile long, half a mile wide, and fifty or so feet high. The shape characteristically resembles the bowl of a spoon turned bottom-side up, with the handle end pointing downstream of the flow of the ice that made it.

Some of the most picturesque bodies left by melting ice are the long, winding ridges of eskers. Since their internal structure is more or less stratified, eskers are thought to be the bottom deposits of streams that flowed through tunnels in or under the ice, or perhaps in open channels on top of it. As the ice melted, the stream beds were let down intact on top of the ground. Some eskers can be traced for miles, even up and over low divides between valleys. They are found in a number of places in the Middle West, as for example south of Lansing, Michigan, and in the central and far northeastern parts of Minnesota.

Most widespread of the elaborations on a drift sheet are the ridgelike end moraines. Although a large part of the rocks, soil, and other debris gathered up by moving ice is laid down along the way as a sheet of ground moraine, a good proportion remains in the glacial conveyor belt all the way to its melting edge. If the end of the conveyor stands in one place for a time, its deposits build up thicker and thicker. On the other hand, when the unloading end is moving either forward or back, there is no local accumulation. This accounts for places where the edge of a drift sheet has no morainal ridge but merely thins out to an indefinite edge. No moraines are built, either, if meltwater is voluminous enough to wash the drift away and beyond as fast as it comes out of the ice.

The situation within an end moraine could be described as total chaos. The material that becomes embedded in the ice is arranged entirely at random, except insofar as it keeps working its way upward along the internal lines of flow. As a result of this movement, the ice surface may become completely covered with drift, even some distance back from the glacial margin. Since the outer part of a receding glacier is more or less falling apart as well as melting rapidly, the drift becomes tumbled and jumbled with pieces of ice, all stirred in with meltwater and forming a wild conglomeration of holes, hummocks, crevasses, streams, puddles, and ponds — in, on, and even under the main body of ice. At such a place it is hard to say just where glacier ends and moraine begins.

As an ice sheet melted away for the last time, its retreat

was very irregular, with many halts and temporary re-advances. The thinning edge usually developed an elaborate outline, bulging into every valley and lowland and curving back around the bases of hills and ridges. Consequently the end moraines that developed are strewn over the country in a looping pattern that can be read almost as a contour map.

Largest and most conspicuous of the end moraines are the huge masses that piled up where two major ice lobes converged, particularly those that moved southward down the troughs of the Great Lakes. Much of the Lower Peninsula of Michigan is covered with such interlobate moraines. From the base of the "Thumb" southwest into Indiana the land is roughened into a knob-and-kettle landscape that makes an attractive setting for many small lakes. There is more such country in southeastern Wisconsin, where the ice lobe in Green Bay grazed the larger one moving down Lake Michigan. Here, in a strip that runs from the Door Peninsula beyond the Illinois line, the hills and hollows are especially sharp and picturesque. Wisconsin has a Kettle Moraine State Park, and a Kettle Moraine Drive has been marked out over country roads through part of the region.

At the other end of the morainal scale are the faint swells that form concentric arcs on the flat land southwest of Lake Erie, Lake Michigan, and Saginaw Bay. These are much more conspicuous on a geological map than they are on the ground. Even a road map shows where they are, because there are still some old roads that follow the slightly higher land of the moraines rather than the square grid of section lines. Rivers follow the shallow sags between the slight swells, and as Fenneman has observed of the moraines, "Their minor role in topography is in strange contrast with their absolute control of drainage."

When a glacier was in the process of retreating, the large mass of rapidly melting ice produced an enormous volume of water. On the rough surface of a moraine this would be channeled into fast-running streams, where the turbulent water became heavily loaded with all kinds of sediment, even sizable rocks. But as it poured down from the moraine and washed out onto more level land, the current was sharply checked, and the water began to drop its load.

As the outwash spread out beyond the moraine, it built up into a broadly sloping plain, with here and there a steeper alluvial fan where a stream debouched from the morainal hills. Water that ran off into an existing valley often filled it deeply with the long plug known as a valley train. Sometimes the outwash built up faster than the ice front receded, creeping up over the thinning edge of the ice or over detached blocks of it. When such buried ice at last melted, the overlying debris slumped down into the vacated space, often forming a topography aptly described as a pitted plain.

Once the torrents that deposited it have subsided, an outwash plain or a valley train becomes scored and channeled by a new generation of smaller streams. These carry off merely the rain and snow that fall locally. This, however, is enough to cut sharp gullies in the loose outwash. With the rapid drainage that results, the surface dries quickly and the fine sediment is readily picked up by the wind. The clouds of dust that hang over such large glacial valleys as the Matanuska in Alaska on bright, breezy days show vividly how the wind gathers up fresh silt; and anyone who has traveled on unpaved roads over outwash-covered parts of Alaska or Canada knows at first hand the fine textured, all pervading, wind-susceptible nature of the stuff.

Thus it was with the winds that swept over the land during the centuries while the ice sheets were ebbing. During that time much of the Middle West became blanketed over with wind-laid silt, or loess. Near the Mississippi in Illinois and the Missouri in Kansas, especially on the east or generally downwind side, the loess is over a hundred feet thick. There are other deeply loess-covered areas in the lee of the Illinois River and the Platte. Away from the great rivers the loess blanket is thinner, as well as finer in texture; but over a large part of the Middle West it lies from a few inches to a foot or so deep.

High winds may carry not only silt but also the coarser

particles of sand. Being heavier, sand requires a stronger wind to move it, and it is not transported for such long distances. It commonly appears in the form of dunes shortly downwind from its source, whether that be glacial outwash or the shores of lakes or rivers. There are large tracts of sandy dunes, now more or less bound by vegetation, in the wide, flat Kankakee Valley southeast of Chicago and on the huge delta made by the Sheyenne River in eastern North Dakota, where it flowed into now extinct Lake Agassiz. The large expanse of outwash known as the Anoka Sand Plain, north of Minneapolis, also has its dune fields.

Ice-laid, water-laid, or wind-laid, a drift surface sooner or later acquires a degree of stability, and plants begin to find a foothold on it. The first colonists are undemanding pioneer types that can make a living from raw ground-up rock, devoid as it is of humus and deficient in nitrogen. As generation after generation of plants live out their lives and die and their organic remains slowly accumulate, the upper part of the drift gradually turns into a soil, meager at first, but eventually becoming dark and fertile.

One of the most striking kinds of drift-derived soil is the "gumbo" of Iowa and thereabouts. Its name suggests its slick and sticky nature when it is wet. As it dries, it hardens like concrete. Geologists have modified the common word to the technical term "gumbotil" to specify a kind of fine gray clay that is produced when glacial till becomes thoroughly decomposed.

The conversion of fresh till into gumbotil requires an extremely long time, much longer than the few thousand years since the last ice disappeared, and it never appears on deposits of Wisconsin age. The warm interglacial period before the Wisconsin was much longer, and the Illinoian drift has developed a gumbotil four to six feet deep. Still longer and warmer was the interglacial period before that, for on the ancient Kansan drift gumbotils reach depths as great as twelve feet.

Many samples of gumbotil as well as other types of ancient soil have been preserved as fossils under layers of drift set down from later glaciers; for the thin outer edge of an ice sheet has so little erosive force that it often rides right over the top of such loose stuff as soil without noticeably disturbing it.

In thousands of places meltwater streaming from the diminishing ice sheet found no ready way of escape. Where the slope of the land was toward the ice, water became ponded against its receding face. Sometimes it was drift left by the ice itself that obstructed the runoff. Many of the lakes and ponds that came into existence this way lived out their lives in a few seasons and soon became filled or were drained as new outlet channels were cleared. But the direct descendants of many such glacier-spawned lakes remain to this day.

Such are the Great Lakes, whose first small beginnings appeared as the ice fell back for the last time to the north and northeast. As soon as the ice front crossed the watershed into the St. Lawrence drainage basin, water began to accumulate against it, and as it receded along the old river valleys that had now been scoured into large troughs, the Great Lakes came into being.

It was something over sixteen thousand years ago that Lake Chicago began to form near what is now the south end of Lake Michigan. At about the same time Lake Maumee appeared behind the ice withdrawing from the western end of the Lake Erie basin. For both these incipient lakes the lowest available outlets led toward the southwest. Lake Chicago drained through the DesPlaines River into the Illinois and on to the Mississippi. Lake Maumee overflowed through a gap in a moraine at Fort Wayne and on into the Wabash.

For the next several thousand years the geography of the Lakes was in a constant state of flux as the ice margin backed and filled. The entire story is exceedingly complicated, but some of the highlights are easy enough to read in the modern landscape.

Until quite recently Lake Chicago and its descendant, Lake Michigan, continued to flow out through the Illinois River. For much of the time the various ancestral stages of Lake Erie were confluent with Lake Huron, their combined overflow sometimes escaping via the Grand River across Michigan into Lake Chicago. Lake Ontario came into existence haltingly and late; for a long time it emptied through the Mohawk Valley and thence down the Hudson.

The ancestor of Lake Superior at first drained westward from a point near Duluth and then via the Kettle River to the St. Croix. Later and for a much longer time the lake outlet was through the Brule in northern Wisconsin, and thence into the St. Croix. For a while it may also have overflowed the lowest part of the Upper Michigan Peninsula into Lake Michigan.

Finally there came a time about four thousand years ago, after Erie and Ontario had become essentially as they are now, when the three upper Lakes were joined into one large body of water. This bears the name of Glacial Lake Nipissing. Nipissing's shores were close to the present shores of the various Lakes, but the water stood considerably higher, high enough to cover the entire Mackinac–Sault Ste. Marie region and to engulf the much smaller modern Lake Nipissing. Through nearly all of its history, this large body of water had three widely separated outlets. One near North Bay, Ontario, led down the Ottawa River to the St. Lawrence, another down the St. Clair River into Lake Erie. The third was the old outlet at Chicago.

Several things happened that transformed Glacial Lake Nipissing into the modern Lakes. For one thing, as the huge mass of ice melted and flowed back into the ocean, the earth, which had long sagged under this tremendous weight, gradually rose again, tilting up on the northeast where the ice had been centered. This lifted the northern side of the lake so much that the sill of the North Bay outlet ran dry, while Lake Superior became separated from Lake Huron except for the rough and rocky channel of the St. Mary's River.

Meanwhile there were changes at the two southern outlets. The spillway at Chicago lay over bedrock and remained fairly stable. But the St. Clair outlet lay on loose glacial drift. There the channel was rapidly lowered, and along with it the level of the entire lake behind it. It was not long before the spillway at Chicago was also left dry, although not very high. In fact, it is still so low that within historical times the continental divide separating the drainage basins of the St. Lawrence and the Mississippi, those mighty rivers, lay through a flat, swampy region at an elevation all of ten feet above the surface of Lake Michigan. In seasons of high water the Indians could push their canoes across it without taking the trouble to make a portage. Even in 1848, before the days of the steam shovel, it was a simple matter to dig a navigation canal through this continental divide.

In their earlier stages, all the Great Lakes covered much larger areas than they do now. The old lake bottoms now spread broad and smooth as tabletops, making the plains southwest of Lake Michigan, Lake Erie, and Saginaw Bay, for instance, surely some of the world's flattest country.

Some of the till plains that were never covered with water are very nearly as smooth. From Ohio to Illinois, even the low moraines have little effect on the general flatness. However, a little digging below the surface quickly shows the difference between randomly ice-laid till and the clearly layered sediments of an old lake bottom.

Such differences are important clues to the glacial history of a place; for it is from the fitting together of countless local histories that the whole drama of the Ice Age has unfolded before us.

An example of this is the famous Two Creeks forest bed that appears in the bluffs along Lake Michigan not far north of Manitowoc, Wisconsin. Here, a number of feet below the surface, a layer of gray till records the passing of an ice sheet. Over this lies a bed of clay several feet thick and interbedded here and there with silt and sand. This was laid down on the bottom of Lake Chicago as it grew northward behind the retreating edge of the same ice that had set down the underlying till. Later, as the ice retreated still farther, it uncovered an outlet far to the northeast, and the water drained entirely away from the Two Creeks region.

On the now dry lake bottom a forest of spruce, jack pine, and birch developed. Remains of some of the ancient trees still stand in place today, with their roots reaching down into the clay. There are also remains of such creatures as molluscs, mites, and wood-boring insects, and of mosses and fungi. The forest grew and thrived long enough to build up a layer of humus and make a real if shallow soil out of the top part of the old clay bed. Growth rings in the oldest tree that has been found show it lived for 142 years. Then the ice reversed itself and began to creep back again, and once more the water was ponded against it, flooding and killing the forest. On the bottom of the restored lake another layer of sediment accumulated, filling in around the trunks of the dead but still standing trees. Then the ice itself arrived and spread another layer of till on top of the latest lake bottom clay.

It is dramatically clear that the front of the oncoming ice moved over the drowned forest like a bulldozer, knocking over the trees, breaking them off where they protruded through the soil surface, smashing some of them to bits, and carrying the fragments along as it moved on. For in this till are embedded the broken-off remnants of tree trunks whose roots are still in place in the soil below. At least one tree has been found with its upper part knocked over and shredded but not quite severed from the roots.

Wood from this ancient forest was one of the very first geological materials to be dated by radiocarbon means. Other samples have been tested since then, and all of them indicate that the Two Creeks forest was destroyed about 11,850 years ago.

The ice that demolished this forest was the last that ever pushed so far south. Not far beyond, it came to a halt and

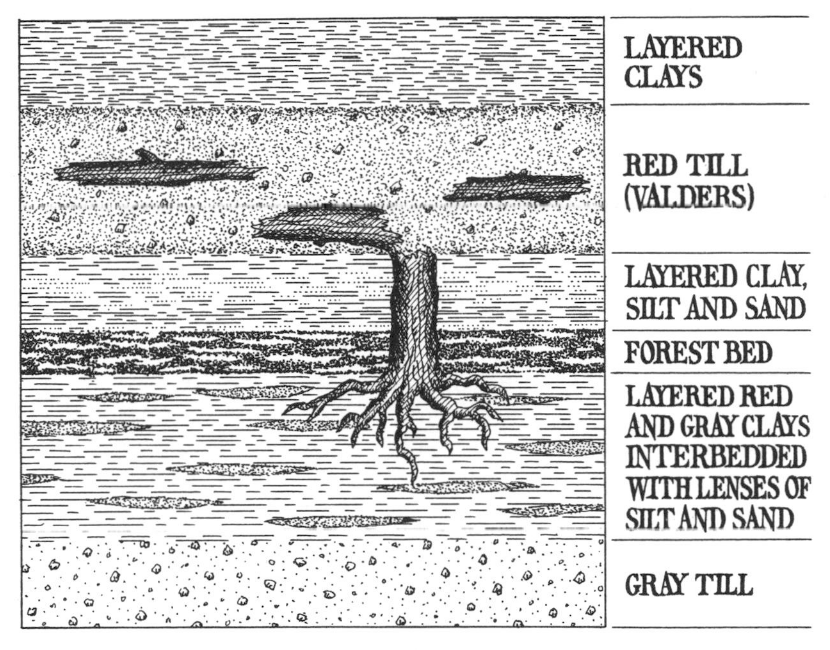

CROSS-SECTION OF TWO CREEKS FOREST BED

went into its final retreat. As the ice withdrew for the last time, once more it impounded a lake that laid a final topping of clay over the latest till, and upon this grows the vegetation of today.

The earlier, higher stages of the Lakes also left their marks around the shores in the form of beaches and bluffs, deltas, bars, and sand spits. It is not very hard for a sharp eye to recognize such old, abandoned wave-wrought features.

Many substantial fossil beaches are strewn over the old lake-bottom plains south and west of Lake Erie and Lake Huron. Some of them are quite easy to find because the oldest roads took advantage of the sandy ridges as ways across the swampy flats. The road that became U.S. 20 east of Cleveland began as an Indian trail that followed an old beach ridge. West of Cleveland the road from Findlay to Delphos and on to Fort Wayne, Indiana, follows a faint ridge along the highest and earliest shore of Lake Maumee. For much of this distance the eye can just detect the difference between the slightly higher, ever so faintly undulating till plain to the south and the slightly lower, perfectly flat lake-bottom plain to the north.

A large, long-lived lake like Glacial Lake Nipissing left a well-developed shoreline that clearly separates the old lake bed from the stream-carved land around it. Jack Hough, one of the leading students of the Lakes, says, "The Nipissing beaches of the northern Huron, northern Michigan, and southeastern Superior basins are among the strongest and most spectacular features of any age found in the Great Lakes region."

The more southerly of the Nipissing beaches stand twenty-five feet above the present surface of the water. But when they are traced northward and northeastward their elevation begins to rise. Near Traverse City, Michigan, and near Duluth they are still at the twenty-five-foot level; but at the Sault and on Isle Royale they rise to forty-five feet, and at North Bay and along the northeast shore of Lake Superior the Nipissing beaches stand ninety-five feet above the present surface of the water. Such facts provide the evidence that the earth's surface has risen in the past few thousand years in a kind of rebound from its sagging under the weight of the ice.

Ever since the drop in water level that ended the Nipissing stage, Lake Huron and Lake Michigan have stood at their present heights. This period of 2500 years appears to be a record for the whole 16,000-year Lake history, as no other Lake stage is believed to have lasted more than a few hundred years. The other Lakes are still rising against their southwestern shores as the land continues to rise at the northeast where the outlets lie, although the rate of rise appears to be diminishing.

Through all the early history of the Great Lakes, while ice

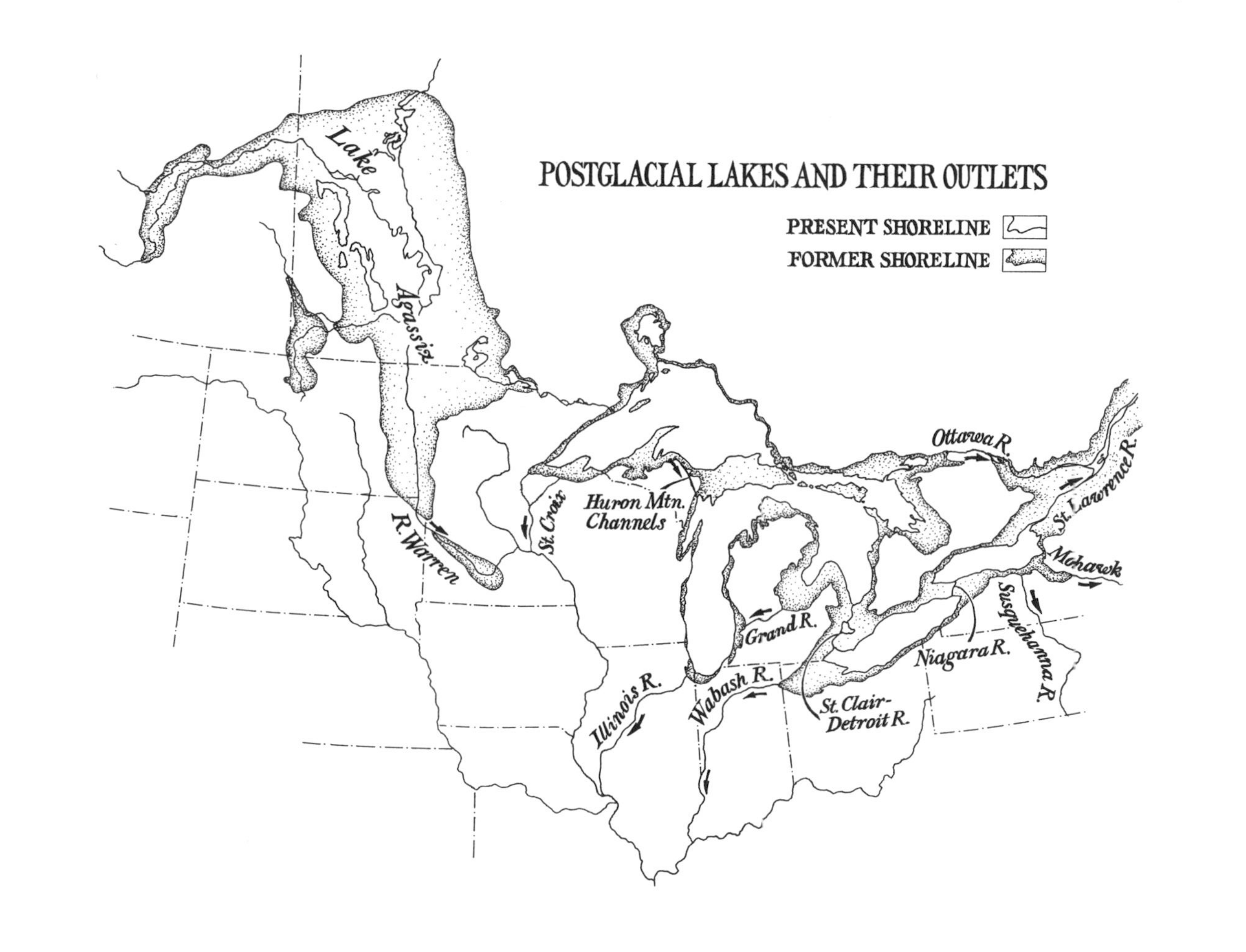
POSTGLACIAL LAKES AND THEIR OUTLETS
PRESENT SHORELINE
FORMER SHORELINE
Lake Agassiz
R. Warren
St. Croix
Huron Mtn. Channels
Illinois R.
Wabash R.
Grand R.
St. Clair-Detroit R.
Niagara R.
Susquehanna R.
Mohawk
St. Lawrence R.
Ottawa R.

still covered the drainage ways to the northeast, the overflow from the young lakes ran off toward the south and into the Mississippi by way of a series of large rivers. Reading from east to west, these were the Wabash, the Illinois, the Rock, and the St. Croix. There were also the Grand River of lower Michigan and the Minnesota that drained Glacial Lake Agassiz.

Since these rivers carried the runoff from thousands of square miles of melting ice, they must have had a tremendous volume of flow. Indeed, their remaining channels show clearly the marks of fast-moving water, which probably filled the present valleys right to the brim, and they are commonly spoken of as sluiceways. The rivers that now flow over the floors of the old channels bear little relation to those mighty torrents.

The most striking thing about the sluiceways is their shape. The land around them is generally level, and the old channels cut abruptly into it in the form of wide, U-shaped troughs with remarkably steep sides and flat bottoms. The sides are much straighter than those of ordinary river valleys, and where bedrock is exposed on the side walls or on the valley floor, it has a pared-off, water-worn look. The whole thing is so typical that one could be forgiven for saying that when you've seen one sluiceway, you've seen them all.

The Glacial Grand River was larger but shorter than its present descendant. It began somewhere in what is now a swampy maze near the village of Maple Rapids and followed the course of the present Maple River into the Grand. Below Grand Rapids the channel widens out, and a little beyond that the glacial river emptied into the lake.

The Wabash in preglacial times was a rather minor tributary of the Teays. Its present form began with the birth of Lake Maumee in the western end of the Erie basin. When the rising waters of the young lake found an outlet through the morainal rim at the present site of Fort Wayne, the new river picked a course through a succession of gaps in the concentric end moraines and finally found its way into the upper reaches of the preglacial Wabash. Through the many long years while the Wabash was swollen with melt-water its upper valley developed into a typical sluiceway.

Later, when the lake level fell, the outlet into the Wabash ran dry and the great river was drastically diminished. What is left of its upper part is only a small stream flowing on the bottom of a large trough. East from the old outlet the present Maumee River follows the southern rim of the former lake for a few miles where it narrowed into the sluiceway. This river is a strictly recent affair that flows in a shallow little valley over the floor of Glacial Lake Maumee to the present Lake Erie.

Just before the earliest Great Lakes came into existence there was a time when the Kankakee River carried off all the water from the ice that lay melting in the Erie and Saginaw and Michigan basins combined. The upper part of the Kankakee, where it crosses southern Michigan and northern Indiana, was probably more of an anastomosing maze than a clearly defined river, for the flood of water had to find its way over the chaotically jumbled surface of a newly laid interlobate moraine. Its course across Indiana was a wide swamp until only a few decades ago, when ditching and draining were undertaken. A number of rather wild marshy tracts remain even today.

Just south of the still unborn Lake Chicago the Kankakee flowed into a broad, shallow lake that was fed also by the DesPlaines. Near the town of Marseilles the lake water swept into the Illinois River through a narrow gap in a massive end moraine that parallels the end of modern Lake Michigan.

This reach of the valley shows all the characteristics of a typical giant sluiceway. The flat valley floor has been cut down into bedrock, and in the straight, vertical side walls the horizontal strata of rock show clearly. Illini State Park lies on the floor of the trough, and a small creek there cuts through the shallow soil to expose much bare bedrock. It takes only a small effort of imagination to refill this great sluiceway up to the brim with fast-moving icewater.

The Rock River had its day as a sluiceway for a time when it served as the outlet for Glacial Lake Oshkosh, a small remnant of which survives as Lake Winnebago. Although this is the largest lake in Wisconsin and lies in a valley whose rock floor is as much as three hundred feet below the land surface, it is all of twenty feet deep! A little farther south the great Horicon Marsh lies in a low place on top of the outwash that now fills this former sluiceway.

The Wisconsin River was never a real outlet sluiceway, but it was much affected by the melting ice. In preglacial times it flowed through and across the Baraboo Hills instead of skirting around their eastern end as it does now. This little east-west range straddles the glacial border, where ice advancing from the east toward the Driftless Area came to a halt. The end moraine that marks the farthest advance of the ice runs directly across the range, plugging at both ends the old gorge by which the Wisconsin had traversed it. The moraine did not entirely fill the gorge, however, and a small sealed-off hollow was left in the middle. Outwash from the surrounding area has partly filled the hollow, but it is still deep enough to hold a small, undrained body of water known to thousands of swimmers and skaters as Devil's Lake.

The plugging of the old gorge dammed the river, backing it up as far north as Wisconsin Rapids to form what is known as Lake Wisconsin. When in time the lake drained, the river picked its way more or less at random over the irregular top of the sandy lake-bottom flats.* Eventually the river found a new way around the east end of the range.

North of Baraboo, as the river worked rapidly down into the fresh sediment, it soon reached the top of a buried ridge of soft Cambrian sandstone. Into this rock it has cut a narrow gorge that is known as the Dalles. The steep,

* This area is Aldo Leopold's "Sand County."

cleanly cut walls of the gorge show beautifully the bedding and cross-bedding of the pale, tawny sands deposited so many aeons ago in the Paleozoic seas.

The present course of the river brings it, at the town of Portage, within a mile or two of the Fox River. The two rivers flow in exactly opposite directions, however, the Fox going northeast into Green Bay. The divide between them is inconsequentially low, and it was a historic crossing place between the Great Lakes and the Mississippi in the days when travel in the region was by canoe.

When Glacial Lake Duluth lay in the western end of the Lake Superior basin, all the overflow from it came down to the Mississippi by way of the St. Croix. At that time the lake margin stood well above the present lake and several miles inland. Part of the Skyline Drive in the city of Duluth follows the abandoned shore of this earliest lake, which now stands 560 feet above the water.

The earliest and highest outlet for Lake Duluth was by way of the Kettle River. Later and for a much longer time it led through the modern Brule River in Wisconsin and over the present divide into the headwaters of the St. Croix, producing the typical large trough of a sluiceway. When other, lower outlets opened to the east, this channel in its turn was abandoned and it, too, became a flat, marshy watershed that provided an easy canoe crossing. The place is now part of the Brule River State Forest.

The Mississippi has been one of the master streams of the great Central Lowland of North America since long before the Ice Age. Although it was never so drastically changed by the ice sheets as the Teays, several parts of it were appreciably affected.

The uppermost part of the great river flows through deeply drift-covered country and its exact preglacial route is not certain. North of Minneapolis it crosses the Anoka sand plain, a tract of outwash ten to twenty miles wide that was laid down in a temporary ice-dammed lake.

From St. Paul to Clinton, Iowa, the river appears to follow

its preglacial course. Through much of this reach it flows in a large trench cut deep into the Paleozoic rocks along the edge of the Driftless Area. As in other places where relatively silt-free tributaries came in from unglaciated country, the valley train building up in the main channel rose higher than the floors of side valleys, damming the tributary rivers; and from the Mississippi trench many small, silty lake-bottom flats reach up into the branch valleys.

One river that did bring in a substantial load of sediment was the Chippewa, then as now a fast-moving stream that descended through steep country. Where it flows into the Mississippi, the Chippewa has built a flat and marshy delta that is massive enough to dam the great river and form Lake Pepin.

Below Clinton the Mississippi suffered many vicissitudes during the Ice Age. In preglacial times it continued southeastward from the bend below Clinton to join the now extinct Teays at a point near Lincoln, Illinois; but as the various ice sheets came and went it was repeatedly shunted from one course to another. Most of the old channels are obscured by glacial drift, but one of them can still be traced across southeastern Iowa through a course that includes parts of the lower valleys of the Wapsipinicon, the Cedar, and the Skunk. Its present valley in that reach is a narrow channel cut in rock, attesting to its relative youth.

While the last waning ice sheet was changing the geography of the Great Lakes region, far to the northwest a much larger lake came into being. This was Glacial Lake Agassiz. Although much the greater part of it lay northward in Canada, where Lake Winnipeg, Lake of the Woods, and a horde of satellite lakes form what is left of it, Lake Agassiz also covered a large expanse in the Red River Valley in North Dakota and Minnesota. Here Red Lake remains as a last outlying fragment, along with a number of smaller and equally shallow lakes and the huge northern muskeg of the Koochiching lands that is sometimes, in a gross understatement, called the Big Bog.

Like the Great Lakes, Lake Agassiz had its beginning when the ice front retreated over a height of land and began to pull back into a northward-draining lowland. The place of origin is now occupied by Lake Traverse, at the extreme northeastern corner of South Dakota. Water rising against the ice soon found an outlet toward the south near the town of Brown's Valley, Minnesota. There it poured through a spillway into the valley of the present Minnesota River in a huge torrent known as the River Warren.

The site of the outlet now forms a low watershed between the south-flowing Minnesota River and the north-flowing Red River. On a map, relations are complicated by Lake Traverse, which lies in the tapering end of old Lake Agassiz, and Big Stone Lake lying in the head of River Warren.

On the ground things are considerably clearer. This is prairie country and open farmland, and from the road between Wheaton and Brown's Valley there is a clear view down the narrowing trough where Lake Agassiz swept toward its outlet. The trough is deep here, and as it narrows into a sluiceway, its walls become steeper and straighter where the current, running brim-full, gathered momentum, and far at the end appears the head of the River Warren.

Southward in the river trench lies Big Stone Lake, dammed at its outlet by a small delta. The road along the lake now runs close to the shore and now swings up onto the rim of the valley. Gravel pits dug into the boulder-strewn valley side show that in some places the sluiceway is cut directly into massive, stony till. Other excavations show the typical stratification of water-laid sediments.

Below the end of Big Stone Lake the sluiceway widens out to a broad expanse covered with outwash. The bedrock floor is not far below the surface, however, for strewn over the plain are the protruding tops of smoothly polished granite bosses. In this country where little bedrock is exposed, these granite outcrops are extensively quarried. Above Redwood Falls more of the bedrock channel floor shows through the covering of outwash, revealing its smoothly rounded contours, with here and there grooves and potholes worn by rocks that were tumbled around by the fast running current. Just so might the bed of the Niagara River look if it were to be drained.

The waters of the River Warren flowed into the Mississippi near St. Paul. For five or six miles downstream from this, the greatly augmented Mississippi cut a deep gorge. Later, when the glacial torrent had shrunk into the modern Minnesota River, the Mississippi here lost much of its force, and the gorge filled to a depth of fifty feet with outwash. Over this the river now flows in a sluggish and swampy fashion.

When Lake Agassiz first began to overflow into the River Warren, the outlet lay over loose glacial drift, and for a time the lake level dropped fairly fast. But when the channel bottom reached solid rock, downcutting was sharply checked and the lake became stabilized for a long time.

Meanwhile Lake Agassiz continued to grow behind the retreating ice front. At its maximum the lake was larger than all the Great Lakes combined and reached north beyond the Saskatchewan and Nelson Rivers in Manitoba, a distance of seven hundred miles from the outlet at Brown's Valley. The narrow southern arm of it was eighty miles wide at Fargo, North Dakota. Along the Canadian border it spread thirty-five miles west of the Red River in one direction and over two hundred miles east, beyond Rainy Lake, in the other. Where the present Red River crosses the border Lake Agassiz stood over four hundred feet deep.

On the bottom of this vast lake, sediment accumulated to a great depth, filling the preglacial valley to an absolute flatness. The lake floor that forms the present land surface has scarcely finished draining, and from a geological point of view, streams have not even begun to erode it.

Along the shore of such a large expanse of water, breaking waves had a strong effect. During the hundreds of years while the lake level stood constant, a massive sandy beach developed that is still clearly visible more than ten thousand years later. This beach has been traced almost continuously around the lake, even far north into the Canadian wilds. In North Dakota it runs more or less parallel to the Red River and forty miles west of it. In Minnesota it swings east from Brown's Valley, then north, until near Crookston it veers off sharply to skirt the southern edge of Red Lake on its irregular way toward somewhere east of International Falls.

Lake Agassiz practically disappeared during the glacial retreat that is marked by the Two Creeks forest. When later the ice returned, Lake Agassiz II developed. It, too, stood at a constant level long enough to develop a strongly marked shoreline. This more or less parallels the earlier shore but lies some eighty feet lower and, in this level country, a number of miles away. In northwestern Minnesota, State Highway 11, the old Sand Ridge Trail, follows fairly closely along the ancient beach ridge for the forty miles between the towns of Karlstad and Roseau.

During the life of Lake Agassiz and while the long beaches were forming, several of the rivers flowing into it constructed large deltas in the edge of the lake. Greatest of all is the delta of the Sheyenne in North Dakota, an enormous pile that covers an area of eight hundred square miles. Its top is a wide, level sand plain, in places blown into a dunish topography. It is even more sparsely settled than most of that part of the world, and a part of it is now the Sheyenne National Grassland. State Highway 46 skirts the edge of it and is a fine place to hunt fossil beaches.

Many other ice-dammed lakes existed for a while in this part of the country. In South Dakota there was Lake Dakota, which flattened the floor of the James River Valley.* Farther west and spanning the Canadian boundary was Lake Souris, where the Souris River now flows through a landscape much like that of the Red River Valley.

* Local inhabitants claim that the James is the longest unnavigable river in the world.

More abundant are the surviving lakes that lie in the hollows of moraine surfaces and other hollows formed in the adjoining outwash by late-melting bodies of buried ice. In most of the Middle West rainfall is great enough to keep such concavities filled to overflowing, so that water moves from one to another and eventually into a river that leads on to the ocean. But in the morainal country of North Dakota rainfall is rather sparse, and the water in lakes and ponds does not usually overflow the divides between them, escaping only by percolation underground or evaporation into the air. One area of 3500 square miles drains into Devil's Lake, which has no surface outlet, although it comes within ten miles of the Sheyenne River, and in the short span of postglacial time has already become brackish. There are other smaller areas with no external drainage, and some whole counties in the Dakotas have no perennially running streams at all.

The ten thousand years since the end of the Ice Age have had little effect on the glacial remodeling of the continent. Given time, however, erosion by running water will shape a geologically more normal landscape. Lakes will disappear through a combination of filling at the bottom and draining from the top as their outlet channels become deeper. Already many small lakes have been shallowed into swamps or filled entirely.

Even the Great Lakes will go in time. Lake Erie is doomed to extinction in the not so very long run. If Niagara Falls were allowed to recede at its present natural rate, it would reach the end of Lake Erie in a mere twenty-five thousand years. Considering that the lake is scarcely more than fifty feet deep except for a small hole in the eastern part, tapping by the Falls would reduce it quickly to a mere valley traversed by a large river. This would give a great spur to downcutting in the St. Clair and Detroit Rivers and rapidly lower the level of Lakes Huron and Michigan.

Change impends also on the southwest side of all the Lakes, where the watershed is not very many miles from the lake shore. It would not take much headward erosion of the valleys leading toward the Mississippi to tap the basins of the Lakes. Everything considered, Lake Superior in its very deep, hard-rock basin seems destined to last longest. But before it has disappeared the ice may come back again; for we really do not know whether we have come out at the end of the Ice Age or whether our times are a major or perhaps only a minor interglacial episode.

Divisions of Pleistocene Time

(Glacial periods are believed to have lasted about 50,000 years, interglacial periods much longer.)

Glacial periods	*Interglacial periods*	*Events*
	Recent: The past 12,000 years.	Slight weathering of Wisconsin till. Soils only weakly developed. Scant modification of glacier-formed topography.
WISCONSIN 50,000–12,000 years before the present		Till deposited that forms the present surface over a wide area.
	Sangamon	Weathering of Illinoian till to form gumbotil 4 to 6 feet deep. Extensive and still well-preserved soil developed. Deposits of peat, loess, volcanic ash. Pollen record indicates a climate similar to the present.
ILLINOIAN		Till deposited generally thinner than earlier tills. Moraine topography still remains on broad, poorly drained areas; elsewhere eroded away.

Glacial periods	*Interglacial periods*	*Events*
	Yarmouth	Weathering of Kansan till to form gumbotil to depths up to 12 feet. The longest and warmest of interglacial periods. Peat deposits indicate a climate slightly warmer and drier than the present.
KANSAN		Till desposited 40 to 150 feet thick. No moraine topography now left — all eroded away.
	Aftonian	Weathering of top part of Nebraskan till to form gumbotil 8 to 9 feet deep. Loess deposited. Animal remains indicate a cool temperate climate.
NEBRASKAN		Till deposited to a thickness of 200 feet; rich in clay. Concealed everywhere by later till deposits. Exposed only in stream valleys, wells, etc.

Worldwide cooling of climates and increased precipitation introduce the start of the PLEISTOCENE somewhere around 1 million years ago, possibly earlier.

4. Reclothing the Land: The Primeval Green Mantle

WHEN THE FIRST Europeans looked upon the country that is now the Middle West, the land was clothed with verdure, full and abundant everywhere, although varying markedly from place to place. In the east lay broadleaved deciduous forest, deep and green in summer but austerely bare after the autumn fall of leaves. Northward, around the upper Great Lakes, a proportion of evergreens softened the bleakness of the long, cold winters. Toward the west the trees gave way to vast expanses of open prairie where the grass grew shoulder high in a kind of luxuriance the newcomers had never before seen.

Yet it was not so long since the land had been freed from a thick cover of perpetual ice, and the plants and animals that lived there were not many generations removed from their first immigrant ancestors. It is only in the past few decades that we have learned something of how the bare till sheets, moraines, and outwash plains were transformed into the landscape we know.

A large part of what we know derives from clues hidden in peat bogs and underwater sediment; for all the countless ponds, lakes, marshes, bogs, and swamps so characteristic of a recently glaciated land are storehouses where samples of things living nearby are filed away over the years. There are deposited the bodies of plants and animals that have lived and died in the wetlands, or near enough to be washed or blown into them, and laid down season by season in precise chronological order.

In the airless debris at the bottom of a still pond or swamp, decay proceeds slowly and incompletely, and organic remains may accumulate to a great thickness. Some of the once living tissues become fragmented and decomposed beyond recognition, but in some Ice-Age peat deposits, even the soft parts of leaves are so little changed that one must be careful not to confuse remnants from ancient times with matted leaves from last spring's floods.

Just such a bed of well-preserved peat came to light in downtown Minneapolis about 1930 when an excavation was being dug for a new building. When chunks of the buried peat were washed free of sand and other fine material, some quite recognizable remains of once living things were retrieved. Among them were wads of virtually intact moss, shells of shellfish, lime-depositing algae, and seeds and cones and bits of bark from tamarack, spruce, pine, and birch. There were beetle wing covers and much-battered pieces of grass seedstalks, looking as though they had been knocked around a good bit before they finally came to rest. There were scraps of charred wood, showing that fires are not a phenomenon only of recent times. The whole assemblage looked like what one might find today in a tamarack–black spruce bog surrounded by an upland forest of white spruce, fir, pine, and birch.

Peat from the forest destroyed by the ice at Two Creeks is very similar, with remnants of spruce, jack pine, and birch, along with mosses and other small plants that grew on the woodland floor. In the course of various excavations and well diggings, peat from this forest has been found and identified by its radiocarbon age over a sizable area of Wisconsin — a thing rarely possible to do.

Other peat beds show that forests of northern type once grew far south of their present range. One that was found near the Mississippi in Illinois contained an abundance of tamarack branches, some of them as much as four inches thick, along with large numbers of well-preserved leaves of both tamarack and spruce. This was a very ancient forest, for over the peat there were deep layers of outwash from each of the last three ice sheets. Six feet of it were from the Kansan and twelve feet from the Illinoian, with a deep soil developed on each one during the long interglacial times. Over the second soil was a thick layer of loess deposited by great dust storms while the Wisconsin ice was approaching. When finally this last ice sheet in its turn melted, the runoff from it added another twenty feet of silt. The top of this forms the soil of today, the third since the demise of the old forest.

Although such deposits may yield a vivid picture of life at a certain time and place, as a guide to the larger past they have serious limitations. Peat that forms in a wet place may contain a fair sample of the plants and some of the animals that lived there; but those living on the drier upland, which is usually much more extensive, are not very likely to be preserved. Their remains would more likely be eaten by animals or insects or eventually reduced by the slow processes of decay to an unidentifiable fine humus.

A more reliable guide to the overall vegetation of times past lies in the minute pollen grains that are produced in huge quantities every spring by all wind-pollinated plants. This living dust is widely distributed over the landscape, and only a small fraction of it ever reaches the appropriate part of another plant where it can fulfill its prime function. The rest settles at random on hill and dale and on lake, swamp, and running river. What ultimately happens to much of it, who can say? But the fraction that falls on a peat bog, or on the surface of quiet water where it drifts eventually to the bottom, will take its place in the accumulating sedimentary file.

For those who would reconstruct the patterns of past vegetation, it is a most convenient fact that the dominant trees,

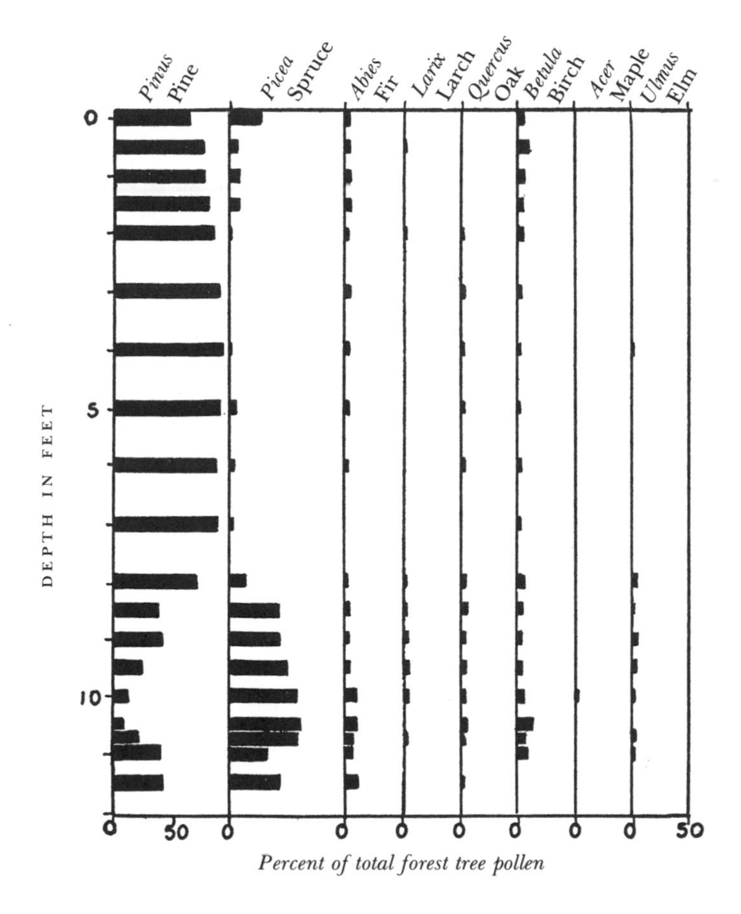

both evergreen and broadleaved, and the grasses and sedges that make up the great bulk of the plant cover in more or less temperate parts of the world are nearly all pollinated by wind. It is also fortunate that pollen grains are readily identifiable, are highly resistant to decay, and may be carried for some distance from their place of origin. On these grounds has been founded the science of pollen analysis, or palynology.

Interpretation of the pollen record has many complexities, and details are still being worked out. Essentially it is based on counting the numbers of pollen grains of the various kinds, using samples taken from a long core bored out of a peat bed or a deposit of lake-bottom mud. From the results a "pollen profile" is constructed. This shows the relative proportions of different kinds of pollen at successive levels in the peat. Presumably it shows also the relative numbers of the plants that produced the pollen at successive times in the past.

For many years the chronology of the past was only relative, and there was no way of knowing with any certainty how many actual years ago any specific event took place or any given deposit was laid down. Cross dating was possible in some cases, using tree rings or clay varves; * but even then there were always pieces of floating chronology that could not be attached with much confidence to any series whose true age was known. This problem was essentially solved in 1949 by the discovery — or invention — of radiocarbon dating by Willard Libby and his colleagues at the University of Chicago. Now it is possible to add a time scale in actual years to our pollen profiles.

In the earlier days, pollen analysis was based entirely on deposits of peat. It was the custom then to bore through the entire thickness of the organic layer, stopping when the coring instrument reached the underlying mineral bed. Such a procedure is likely to miss the first part of the record, since a number of generations of plants may live and die around a newly formed body of water before enough organic matter accumulates to produce a detectable amount of peat. More recent studies include a search through the entire thickness of water-laid material, what-

* Varves are thin layers of water-laid clay of alternating darker and lighter color. In cross section a varved deposit has a banded appearance that somewhat resembles the pattern of tree rings. Each light-dark band or varve is believed to be one year's deposit of clay.

ever its nature. The very earliest deposits consist almost entirely of mineral sediment, and pollen grains may be extremely sparse; but the information they provide is of great interest and importance and quite worth the trouble to find them.

One of the questions that has been pursued by the study of pollen records is, What happened in the regions beyond the glacier during the coldest depths of the Ice Age? How far reaching were the effects of the vast sheets of ice, and where did the plants and animals go for the duration?

Quite apart from the climatic cooling that brought it into being, one would expect that a frozen mass as large as a Pleistocene ice sheet would affect the climate for some distance beyond it. The distance does not seem to have been very great, however, and it probably varied with the latitude. There is evidence, both geological and theoretical, that the lowering of temperature during the Ice Age was progressively less at greater distances from the poles, and that the tropics were little if any cooler than they are at present.

From the pollen record it appears that northern plants migrated southward as refugees from the increasing cold, but not in such proportions as to totally dislodge the resident natives. Northern evergreen forests grew as far south as North Carolina and northern Georgia, but plants of more temperate climates were also present through the cool times. What seems to have happened was a crowding together and a telescoping rather than a bodily migration of entire broad belts of vegetation. Such information as there is suggests that northern Florida was not much affected by the Pleistocene chilling.

Remains of animals are much scarcer than those of plants and much more sporadically distributed, and our knowledge of them is correspondingly limited. However, since the birds and the beasts tend to live wherever their preferred habitats can be found, it seems likely that they followed the vegetation southward. Remains of such northerners as the gray jay and the spruce grouse have been found in deposits from late glacial times in the Appalachians of Virginia.

It is not entirely clear just what conditions were like at the very edge of the ice, either during the height of glaciation or later while the ice was receding. Opinions are divided as to whether or not there was a zone of treeless tundra, and if so, how wide it was. One source of confusion lies in the time relations between the disappearance of the last ice and the beginning of sedimentation in ice-formed depressions. In some cases sediment starts to accumulate in morainal hollows right on top of the ice and some distance back from its edge. Moraine may cover the whole lower end of a retreating valley glacier deeply enough to support a young forest, as one can see in Alaska today. There is no reason to think that the edge of a large ice sheet would be different. On the other hand, some lakes may not appear until the main glacier has retreated far beyond them. These are the ones that form in hollows left by the belated melting of buried blocks of ice. In view of these possibilities, the distinction between glacial and postglacial sediment becomes rather blurred.

In any case, some sort of tundra seems to have existed close to the edge of the ice, for the earliest chapter of the pollen record nearly always shows a high proportion of what is known in the trade as NAP — non-arboreal pollen — mixed with some spruce and fir. NAP comes largely from grasses and sedges and from a number of small heathlike plants, all of which are typical of tundra. The tree pollen mixed with NAP probably represents a parklike array of scattered spruce groves interspersed with larger or smaller tracts of open tundra.

There are also animal remains that indicate a tundra belt of some sort. Bones of several kinds of muskox have been retrieved from glacial deposits in the Middle West. All but one of these are now extinct and there is no way of knowing what kinds of habitats they required; but the surviving

Ovibos moschatus lives strictly on the tundra and cold steppe of the arctic coast and islands of Canada and Greenland. Three skulls of this species have been found: one in Fayette County, Iowa; one in Sioux County, Nebraska; and one in Wabasha County, Minnesota. It is reasoned that where there were muskoxen there must have been tundra, and a certain amount of spruce pollen does not necessarily mean that there were no large expanses of open land to serve as muskox range.

It is quite certain that the first continuous forest that developed in the wake of the ice was a dense stand of spruce and fir. Evidence of a spruce belt appears all across both North America and Europe, and thanks to radiocarbon dating, its progress can be followed northward with time. It occupied the Middle West of this country around twelve thousand years ago. Then as now, spruce meant a distinctly cool, moist climate.

After the spruce, again almost universally, came a forest of pine and oak. These are trees that live in warmer and drier places, and they indicate that the climate had moderated. The pines and oaks eventually reached even farther north than they do today. From this it appears that the climate was once warmer than it is now. The warm period began about seven thousand years ago and lasted for some three thousand years. Then came a time of cooling that reached its extreme around 600 B.C. and shows in the bogs as an increase in the amount of beech and hemlock pollen.

From then on the pollen record is supplemented by archaeological evidence and then more and more widely by historical records. There was a warmer interval around 1000 to 1300 A.D. that coincided with the prosperity of the Norse settlements in Greenland and the cultivation of grapes in England. In the seventeenth and eighteenth centuries there was a "little Ice Age" that caused a general expansion of alpine glaciers in Europe and apparently also in Alaska. The latest detectable trend is toward a slight warming during the past century. This has brought about a rapid retreat of glacial fronts and an advance of trees onto the arctic tundra. It is even faintly discernible in the topmost layer of peat bogs.

It was more than a century ago that the possibility of a postglacial warm period was first suggested. Until then it had always been assumed that the climate has grown steadily warmer since the end of the Ice Age. This novel idea was invoked to explain the existence of certain plants and animals in small areas outside their large, continuous ranges. Detached colonies of prairie plants living in the deciduous forest, or single western species growing in isolated areas east of their general range, are often found in places that seem somewhat less than optimal for them. Such plants could be readily explained as relics from a time of greater warmth and dryness that have managed to hold on in spots where local living conditions still permit them to survive.

In the past few decades thousand of pollen profiles have been studied. Considered together, they show clearly the reality of a postglacial warm, dry period, when many common plants grew farther north and east than they do now. In Wisconsin, for instance, an abundance of both butternut and hickory pollen has been found all through a three- or four-foot thickness of peat in a bog that is fifty miles farther east than butternut now grows and ninety miles farther than hickory. As would be expected, both are lacking from the surface layer of the peat.

A number of cases of apparently misplaced plants can be traced to earlier phases of postglacial history. Such are the spruce, tamarack, sedges, and heaths that commonly grow in undrained, acid bogs throughout the glaciated parts of North America, even in the more temperate regions. These are relics of the northward migration of the spruce forest behind the last retreating ice sheet. Peat bogs characteristically lie in cold, wet situations or frost hollows where there is no ready outlet for cold air or cold water. In such small detached fragments of northern climate, the northern plants linger on, resisting the encroachment of the quite different and more southern vegetation that surrounds them. The white pines and arborvitaes that grow on the north sides of some Indiana cliffs would date back to the pine period that commonly followed the spruce, relics of slightly less antiquity than the bog plants.

As the successive ice sheets came and went, none of them ever quite covered the Driftless Area of southwestern Wisconsin, and this region was always available as a refuge for the hardier plants and animals. The area has a number of peat bogs to provide a pollen record of glacial times, since many streams there became blocked by outwash from the nearby moraines, and so backed up into standing water. It is not very surprising that these bogs record a combination of spruce and fir with pine and hardwoods, including oak. The trees probably never grew entirely intermixed, but were sorted out among small localities that were warmer or cooler, drier or moister, according to their different needs.

The Driftless Area now harbors a very few kinds of plants that are found nowhere else at all. It also has about thirty other species that are found in other places outside the glacial border but never on glaciated land. It is not clear just why these plants have not moved back into the intervening territory that they must once have occupied, as many others have long since done, but their peculiar distribution must somehow be related to the ice sheets.

One group of plants that inhabits the Middle West did not live there in preglacial times. These are coastal plain species, which now grow as far inland as Minnesota. Such plants as beach grass, beach pea, sea rocket, and hudsonia are found on sandy places all around the Great Lakes, as well as on fossil beaches fringing Lake Agassiz and other, lesser postglacial lakes and their outlet floodways. They are common also on the Atlantic coastal plain hundreds of miles to the east; but they are found almost nowhere else.

There is, however, a connecting pathway, although a rather spotty one; for all these plants grow also in dry, sandy places around Lake Ontario and Oneida Lake and along the Mohawk Valley and down the Hudson. This is the route followed by outlet waters from the lower Lakes during a large part of the glacial retreat, and along its edges, seed by seed, the plants have migrated inland. In fact, the existence of this corridor inhabited by typical coastal plain plants is one of the things that helped to convince geologists that the Mohawk-Hudson Valley really did carry a huge volume of water that came from the melting ice.

There are other sandy inland places where one might expect to find plants of the coastal plain. The Sand Hills of Nebraska are one example; but no coastal species at all are found there, in spite of the dune-like nature of the country and its many lakes. Apparently the intervening distance is too great for any such plants to have made their way across it.

Along the Illinois River there are other large sandy tracts where coastal plain plants grow, but these are typical of the Gulf Coast and the lower Mississippi Valley, and they do not grow around the Great Lakes or down the Mohawk-Hudson pathway. Only a short distance separates them from the drainage basin of the Lakes; but such plants as climbing hempseed and the river birch have never made the necessary jump and are not found in the Great Lakes watershed.

While the plants migrated with the coming and going of the ice sheets, there were corresponding shifts among the fauna. These movements had a marked influence in the bird world. Each time the ice advanced and the birds were crowded south, many species were divided into separate populations, some moving southeast, others southwest, still others westward into the mountains.

Within these isolated groups small evolutionary changes took place as they adapted to their new living places. The changes were not great enough to prevent interbreeding, however, and when the ice receded and the birds could once more spread and mingle across a large and continuous range, new hybrid forms appeared. Four times this separation and reunion went on during the four major glaciations. The subspecies found in such birds as the white-crowned sparrow are thought to be the result of this kind of history.

The effect was even more impressive among the northern wood warblers. Just before the final climatic cooling that ushered in the Pleistocene, the warblers inhabited two separate parts of North America. One contingent lived in the boreal evergreen forest, which than lay far north of where it does now. The other group lived in the warmer deciduous forest that covered much of the eastern part of the continent from Canada to Mexico.

With the oncoming of the first ice sheet, the northern population was driven far to the south. Some moved off to the western mountains. Others moved into the southeast-

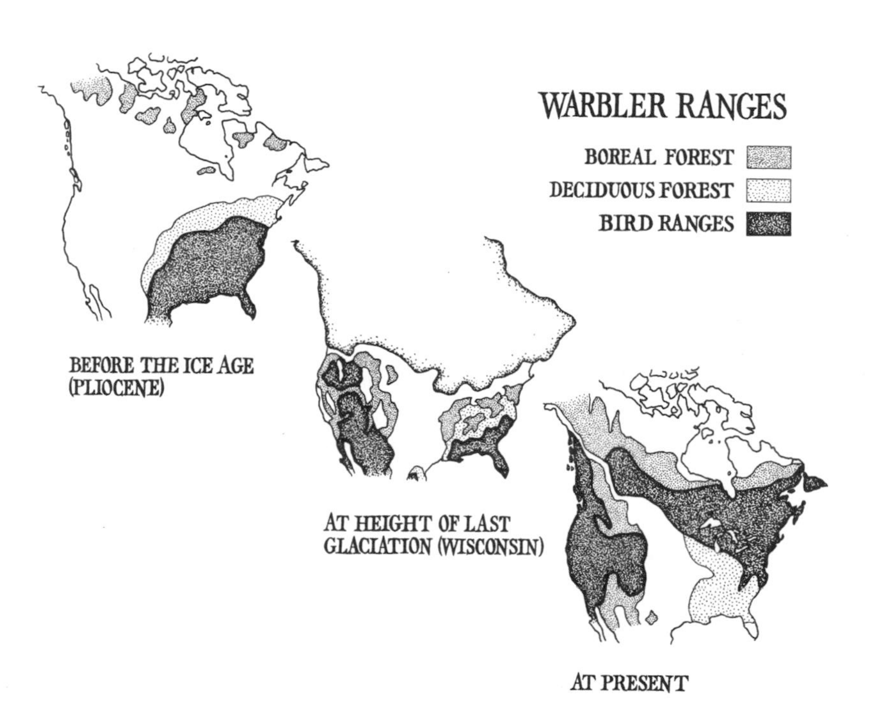

ern forest, where they mingled with the resident species. There the entire mélange was compacted together into a sharply constricted area, and a certain amount of interbreeding took place.

Eventually the ice disappeared, and during the long interglacial time that followed, the hybridized warblers dispersed themselves widely across the continent, adapting themselves in the process to many new kinds of habitat. When again the ice descended, once more the birds were divided into isolated groups, and once more the pressure of natural selection brought about new adaptations. All together, there were four such cycles of compression and expansion, spread over a million years.

Of the forty-six species of wood warbler extant on the continent, perhaps a quarter owe their existence to these events. It has even been suggested that a dozen more may have arisen in this way and already become extinct.

Rather more spectacular than the appearance of new kinds of small birds was the total extinction of many large mammals toward the end of the Pleistocene. Through the later parts of the Ice Age, North America had its own populations of such long-gone creatures as mastodons, ground sloths, llamas, peccaries, camels, horses, giant beavers, and saber-toothed cats. Some of these apparently came as immigrants over the Bering Sea land bridge as recently as 100,000 years ago, and some of them lived here until only a few thousand years ago; the latest of these big animals that has been dated by radiocarbon means is a mastodon found near Tupperville, Ontario, where it met its end about 4300 B.C.

It is still an unanswered question why it should be only large mammals that disappeared, for smaller animals and the entire plant world show no sign of such a fate. The fact that man, the hunter, appeared here at just about the time these animals vanished seems highly significant. It is still being argued whether primitive men could have completely eradicated several kinds of large animals from an entire hemisphere. But Paul Martin has recently constructed a quite plausible explanation of how a newly arrived, migrating, and expanding human population could have achieved such an annihilation in a thousand years.

The disappearance of so many grazers and browsers left a corresponding number of vacant ecological niches — "job opportunities," Martin calls them. The domestic livestock that European man brought with him will not or cannot use much of the available native vegetation. Martin has suggested that we consider reintroducing, after due and careful study of the implications and probable consequences, existing relatives of these extinct beasts that could use this resource and themselves become food for hungry humanity.

It is only in the past two centuries that European man has laid his hand upon the region we now call Middle West. Although every part of the landscape has since come under the ax and the plow, still one might expect somewhere to find undisturbed remnants that would show what the country was like in its primeval state. But so valuable is the land for either timber or farming that truly virgin places are extremely rare and limited in size.

Though the primeval world has all but disappeared, there are ways of finding out what it was like. Even in a highly cultivated area, odd corners that some chance circumstance has spared may hold clues to the original nature of the place. A rocky outcrop or a glacier-dropped boulder may deflect the haying machines enough to let a tree or two grow undisturbed, with a little patch of smaller herbage underneath. Other protected crannies lie in the angles where fences meet at the corners of fields, or along the fenceline itself. In a once forested region, stray patches of arbutus or bunchberry may remain as mementoes of pine and spruce, or spring beauty may linger on where deciduous woods once grew. In a prairie region there will be clumps of prairie grasses, or occasionally the yellow-flowered compass plant whose leaves stand edgewise to the hot noonday sun. Such plants are known to botanists as "relicts."

The exact meaning of such relicts must be interpreted

with caution. Some plants adapt so readily to new situations brought about by man's activities that they migrate far from the places where they originally grew. Others are more conservative and hence more reliable as clues. Many typical prairie plants have spread along the open strips by the edges of roadsides and railroad tracks from the open prairie right through the woods and out again. But a careful study of this matter, made in southern Wisconsin, showed that the compass plant and the picturesquely named rattlesnake master even in 1940 grew only on land that had actually been prairie when the country was first surveyed in 1836.

Another clue to the presettlement vegetation of a place appears in the soil, for the structure and appearance of soil are strongly influenced by the plants that grow on it. The same kind of starting material that under a cover of grass produces a deep, dark, and crumbly soil develops a lighter brown color and less rich texture under deciduous forest, while an evergreen or conifer forest produces a thick surface mat of scarcely decayed needles and twigs, with a sandy layer a few inches down that may be a striking, ashy white.

For a true and reliable picture of the primeval vegetation of a place one might turn to contemporary eyewitness accounts; but the value of these varies enormously with the breadth of the witness's botanical knowledge and the accuracy of his observations. Frederick Clements, a lifelong student of American vegetation, once remarked rather acidly that the myth of the Great American Desert was "largely a tribute to the powers of observation of the layman."

Much more meaningful are the observations of the men who made the original surveys for the General Land Office of the federal government. This work, detailed and systematic, was done before the public domain was officially opened for settlement and before forest and prairie had been much disturbed at all. The purpose of the survey was to lay out a grid by means of which settlers could locate and identify the tracts of land they wanted to claim. To do this, the surveyors ran a set of straight lines, both north-and-south and east-and-west, at one-mile intervals. Every point where two lines intersected was designated a "corner point." All corners were marked on the ground by posts or piles of rocks, and their locations were recorded on a map. Each corner was further identified by noting its distance and compass direction from each of two to four specific blazed trees. The species and size of each "bearing tree," as well as information about other trees observed along the survey lines, were recorded in the surveyors' notes as they worked. Sometimes all the different kinds of trees growing in the area were listed in the order of their relative abundance, along with various general observations on the vegetation, topography, and soil.

The records of these early surveys contain masses of factual data, making them veritable mines of information. The original notes in the surveyors' own handwriting are still on file in the archives of the states and counties, and they have provided the makings of many maps on both large and small scales.

A few years ago a study was made in the Keweenaw region of northern Michigan to see how reliable the old records might be. A number of tracts of land were resurveyed, and as many as possible of the bearing trees and line trees recorded in the original survey were located. In all, 152 section corners were examined, and at 49 of them at least one of the old bearing trees was still standing and could be identified a century later. Even the corner posts were found, sometimes intact, sometimes detectable only as humus-darkened squares outlined in the soil.

The paper reporting this work gives an interesting history of the instructions issued to the surveyors and the methods they used. In it the author notes that the old surveys cannot be used without a certain amount of checking, since "To have subdivided some townships in the time reported, a crew would have had to survey as fast as the men could walk. There are surveys on record where they must have moved even faster." However, "in spite of its shortcomings, the original land survey represents an unusually good job, considering the equipment with which the surveyors had to work and the conditions they encountered. No one can read contemporary descriptions of these surveys . . . without feeling admiration for those who did the work."

One of the conditions they encountered is suggested in this passage from the letter written by Edward Tiffin, then Surveyor General of Ohio, to accompany the report of a land survey of 1815: "The surveyors, who went to survey the military land in Michigan Territory, have been obliged to suspend operations until the country shall have been frozen over so as to bear man and beast."

5. The Ancestral Forest

IF A KNOWLEDGEABLE woodsman of today could enter a magic time machine that would take him back for a walk in the temperate forests of early Tertiary times, 60 million years ago and long before the Ice Age, he would find himself in a world that seemed familiar. He would recognize nearly all the plants in the forest around him, and if he knew the southern Appalachians well, that is probably where he would think he was.

Yet there would be certain differences. For this was a forest on a tremendous scale, and it reached with rather minor variations for thousands of miles in all directions — reached, in fact, all the way around the world over a wide belt of latitude. The zones of climate, each with its corresponding vegetation, lay much farther north than now, and even the northern parts of the present United States and all but the northernmost part of Europe were covered with subtropical forest. The temperate "summergreen" forest lay still farther north, stretching across Alaska, Canada, Greenland, Scandinavia, and most of Russia and Siberia. Beyond that, and as far as the land reached toward the pole, the plants were those of cooler but still temperate climates, and there was no land of perpetual ice and snow. Only in the tropics was the climate like that of today, with the difference, however, that the belt of hot climate was wider. Rather surprisingly, the tropics were al-

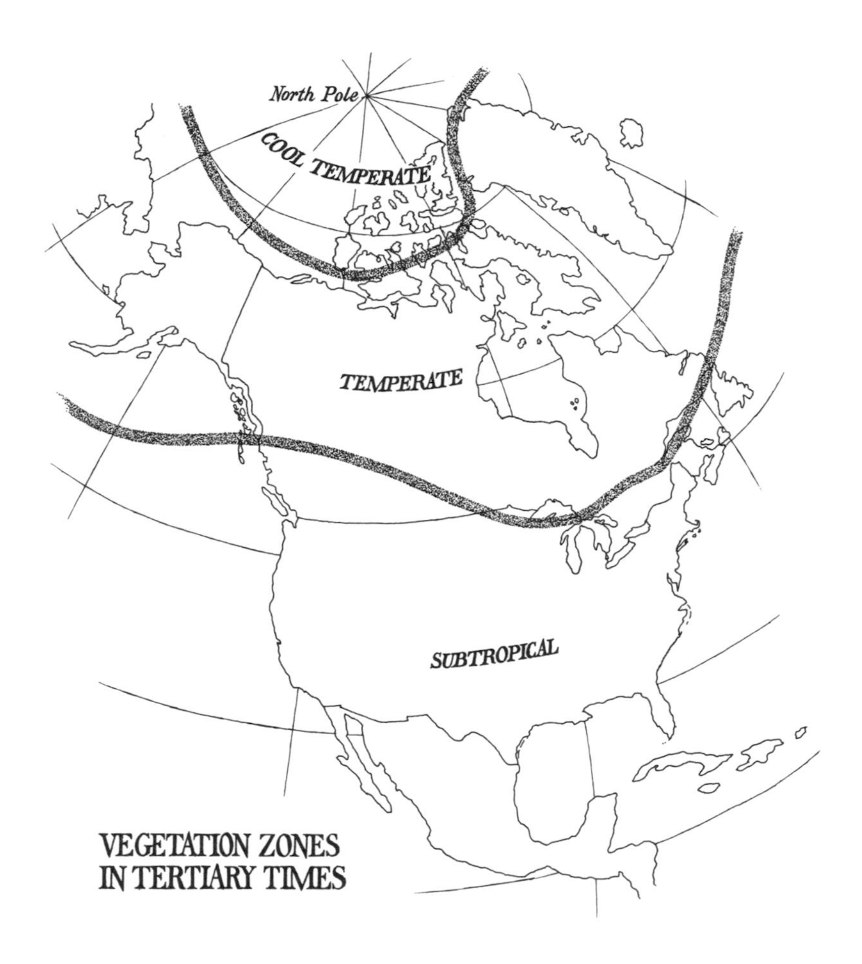

most certainly very little if any hotter than now, and it was only a much more gradual decline of temperature from the equator to the poles that made the great difference from the world of today.

In those remote times the contours of the earth's surface, too, had less contrast, for most of the land was low and flat. With no mountainous barriers to interrupt the free flow of winds from the ocean to regions far inland, the air was universally mild and moist, and rainfall was plentiful everywhere. Nor were there any markedly wet and dry seasons. Even the difference between winter cold and summer heat was relatively small, just as today in regions that have oceanic climates. Such a genial environment offered only moderate challenges in the lives of plants and animals, and a great variety of creatures lived together over a wide range of territory.

In those times the earth's geography also was different from the present. All around the northern hemisphere there were wide belts of more or less continuous land. The shallow floor of Bering Strait between Asia and Alaska stood above water, and North America had not yet become far separated from northern Europe. With no serious obstacles to their migration, the same plants and animals inhabited large areas of North America, Europe, and Asia. To us who are accustomed to the widely varied landscape of the present, the aspect of that ancient world would seem monotonous.

A properly skeptical reader may well wonder how we know all this about a world so remote in time, a world in which man did not yet exist. Primarily, we know all this from the fossil record, the actual physical remains of plants and animals that lived in those times. The plants tell most about conditions, for they are quite directly dependent on the climate, while animals, although they have their own broad climatic requirements, tend to go wherever they find the kind of food and shelter they need, and this is usually determined either directly or indirectly by the general nature of the plant cover.

Even now the fossils of the future are being laid down in almost any kind of place where there is more or less permanent standing water. There you will find layers of matted leaves and probably broken twigs, seeds, and pieces of cones, with an occasional insect shell and perhaps the skeleton of a mouse or the skull of a bird. In the leaf shedding season the great bulk of the debris will consist of fallen leaves from the nearest trees. What is there will be a random mixture of whatever falls, blows, or washes in from the vicinity, all interbedded with larger or smaller amounts of sand and mud.

In the larger, more permanent bodies of water some of this miscellany will become buried more or less intact and at some depth. In its earlier stages of transformation this would turn into peat. To become converted into mineralized fossils, it must be covered with a great thickness of overlying material and infiltrated with mineral solutions for geologically long periods of time.

When aeons later it once more comes to light, such fossilized matter may appear in various states of preservation. Sometimes there is nothing left at all but an imprint or a cast of its external form. Often some of the original matter remains but in a completely flattened state; this is especially common in the case of leaves and twigs. Or nothing at all may be left but the outline of a leaf and the pattern of its veins, sometimes so delicately preserved that they appear to be drawn on the rock surface with a fine-tipped pen. In still other fossils the tissues are virtually unchanged from their original state, and even microscopic internal details can be seen.

Given the fragmentary state of so many fossils, it is a wonder that many of them can ever be assembled and identified. For extremely ancient forms, this may take some doing. Separate jaws or legbones, leaves or seeds may be known for a long time before they are found in organic connection with another known fragment, and reconstruction of a whole organism may take many years.

Things that have counterparts in the living world present less formidable problems. Although many botanists throw up their hands and grumble at someone who brings them a mere scrap of a leaf to be identified, a really good field man can name most of the plants he knows from a leafy twig, and many of them from a single somewhat battered leaf. The student of fossil plants must do even better, using the finer and less obvious details and being able to recognize an isolated leaf by the shape of the teeth on its margin or subtleties of the pattern of veins and veinlets. There are family jokes among botanists about the herbarium man who cannot recognize his plants until they have been pressed and dried; but for a student of fossils a modern specimen pasted flat to an herbarium page is likely to be at least as useful as a live one.

Like their modern counterparts, the plants of ancient times commonly grew together in characteristic groupings or "floras." When the same set of species is found together in place after place, it is reasonable to conclude that they shared a common preference for certain living conditions. Sometimes the structure of the plants offers a clue to what those conditions might have been. Plants with broad, thin leaves, for instance, must always have grown where there was plenty of moisture, and compact dormant buds and growth rings in the wood, in former times even as now, would develop only where there was a dry season or a cold winter. It has probably always been true that where winters are extreme, even woody plants are small or grow close to the ground, like arctic and alpine plants of the present.

For remote ancestors in the distant past, we can draw only very general conclusions; but the forests of early Tertiary times contained many of the same species that are living today. Though the habits of some may have changed, the probability is vanishingly small that all the members of a large and varied flora would have changed in the same direction. So the plants are accepted as keys to climates of the past.

The long cooling that led ultimately to the ice sheets of the Pleistocene is clearly recorded by fossils. As the zone of temperate climate moved slowly southward, a wide swath across the entire continent came to be covered with a forest much like the present redwood forest of northern California.

Well preserved Tertiary fossils are found in many places

in the western part of this country. An especially abundant deposit from the ancient redwood forest has come to light in central Oregon along Bridge Creek, a tributary of the John Day River. This is now a harsh and arid region of sagebrush and juniper; yet the plants whose remains are found in the rocks there are the same kinds that live today near the mild and foggy coast.

In both ancient and modern redwood forests the same four kinds of tree predominate. Among the fossils these are, besides the redwood itself, an oak that is scarcely distinguishable from the modern tanbark oak, an alder much like the common red alder, and a plant that is a near twin of the California laurel. There are also lesser amounts of maple, hazel, and dogwood. Some of the plants of the modern forest, especially shrubs and smaller flowers and ferns, are missing from the fossil bed, however.

This situation some time ago interested Dr. Ralph Chaney, who knows both ancient and modern forests well. In thinking about the resemblances and differences between the two, he pondered the question of what kind of fossil record the present forest might leave and how it might compare with the deposit at Bridge Creek. Since fossils are abundant in the rocks there, and since there is plenty of material in the living forest to compare them with, he decided to put the question to an experimental test.

With an assistant he went to the Bridge Creek fossil bed, near Mitchell, Oregon. There they set about digging out slab after slab of fossil-bearing shale and splitting it into the thinnest possible layers. Every specimen of plant that came to light they identified and tallied according to its kind. In the first day's work they averaged 114 specimens per person per hour. Later they speeded up considerably, and in a total of 80 man-hours, they found and identified 20,611 specimens of 31 different kinds from 98 cubic feet of rock, for an overall average of 210 identifiable fossils per cubic foot!

They found a large number of the present members of

the redwood forest, plus some that now live in the west but not among the redwoods (walnut, sycamore, hackberry), and a few that no longer live in the west at all (basswood, elm, beech). Ferns and other small plants were present but very sparse, much fewer than in the living forest.

For a comparable sample of the present forest, they then went south to Muir Woods, near San Francisco, to look for a place where leaves and other debris were collecting in the way fossils seem to have done. For this they selected the bed of Redwood Creek, an intermittent stream where water lies much of the year in a series of quiet pools. To get truly random samples for comparing with the fossils, they used a foot-square wire frame, which they tossed onto the stream bed where the fallen leaves lay thickest. Where the frame came to rest, they carefully dug out a square-foot slab of the matted debris and identified and counted all the plants in it, just as they had done for the fossils. Allowing for a difference in the amounts of mud and silt, they estimated that a few inches' thickness of this material was very nearly the equivalent of a foot's thickness of the shale. Remarkably, each square foot of pool bottom yielded an average of 200 specimens, very close to the 210 grand average for the fossil bed.

A remaining question was whether the number of specimens bears any consistent relation to the number of plants of the same kind growing in the vicinity. By rather intricate calculations involving the numbers of individuals, the distance from tree to sampling place, and the height of the tree, they showed that the potential fossil deposit does indeed reflect the proportions of trees around it. As for the ferns and other small plants that are scarce among the fossils, although there are plenty of them in the living forest, their leaves are not shed like those of trees but commonly dry up on the stalk and rarely get into the pools and puddles.

The many fossil beds of the John Day River region contain an unusually long and unbroken record of the vegetation of one place. This shows clearly a progressive change from warm to temperate forests and on to the more cold- and drought-tolerant vegetation of today. On the eastern side of the continent the record is much more fragmentary, but it shows the same southward movement of forest zones.

While the climate was becoming slowly colder, other changes affected the shifting vegetation. The old continuous worldwide forest belts were disrupted, in some places by the rising sea, in others by the appearance of mountains. Asia became separated from North America as the broad lowland between Alaska and Siberia was submerged beneath the ocean. The land connection between North America and Europe was broken, forming a number of islands, and on some of these arose masses of high mountains. Henceforth, passage between the continents would be only through regions of Arctic cold, and then only during a few brief intervals of time when sea level fell below the old Bering land bridge. The flora and fauna of Europe and America have long gone their separate ways.

Within North America, too, the great forest belts were broken into parts. At a fairly early date the western part of the forest became separated from the rest as the Rockies began to rise, and in the rain shadow of the new mountains a large dry region appeared. Later the rising Sierras and Cascade Range shut off the western interior from the Pacific Ocean, and the Great Basin, now shielded on all sides from moisture-bearing winds, became a desert. Along the southern Pacific coast drier summers eliminated many plants that had formerly grown there. Only near the northwest coast, where rain is still more or less perpetual, and in the limited area where the summer drought is modified by frequent heavy fogs, do the moderation-loving plants of the ancient Tertiary forest persist in large numbers.

Although the country east of the Rockies is shut off from damp Pacific air, southerly winds in spring and summer bring moisture from over the Gulf of Mexico, and the region is not dry enough to be a desert. But a large part of the center of the continent can no longer support forests and has developed into one of the world's great grasslands. Rainfall is more plentiful in the hilly country of the Ozarks; but although that region, too, is covered with a direct descendant of the ancestral Tertiary forest, it is dryish and supports only a fraction of the species that can grow in a wetter climate.

Another break in the once continuous forest belt was made by a large arm of the Gulf that spread far up the Mississippi lowland. This flooded the present Mississippi Valley as far as southern Illinois. More recently the ice that came down from the north reached practically to the head of the water-filled embayment, and the regions on either side have had separate histories for the past million years.

In the midst of all these changes, one area remained undisturbed, and there a remnant of the once world-encircling temperate forest continued to live on. This was the Appalachian highland, an upland bounded by the flooded Mississippi lowland; the southeastern fall line, where the coastline lay until the fairly recent uplift of the coastal plain; and, on the north, by the glacial border. Within this area the only changes of consequence for over 200 million years have been a moderate fluctuation of temperature and the slow shifting of the land surface by periodic gentle uplift combined with constant erosion.

Ever since biologists began to turn their attention to such things, they have recognized that the southern Appalachian region supports an especially rich and abundant life, both plant and animal. For a temperate forest, it offers ideal growing conditions, with plenty of year-round rain and a generally moist atmosphere, warm summers, and winters that are moderately cold but not very long or severe. The soil in its virgin condition is deep, dark, and fertile, and the land is hilly enough to be well drained but not so steep as to be unstable. In every way it is a country of moderation so far as the life of the forest is concerned.

The richest vegetation of the Appalachians is found in the Cumberland Mountains and in the wooded "coves" of the Smokies — sheltered valleys that indent the sides of the mountains. This bears a strong resemblance to the temperate forest of fifty million years ago. Four out of five kinds of plants of the present forest — all the important trees as well as all the ferns, shrubs, and woody vines — lived there together in early Tertiary times. Even among the small, spring-flowering plants of the forest floor, those that are commonest and most widely spread had almost identical close relatives in the ancient forest. Perhaps the most conspicuous difference is the absence of the once abundant redwoods and ginkgo, plants that grow wild only in very restricted parts of the modern world.

As far back as the eighteenth-century days of the great Linnaeus, when naturalists were combing the remote places of the globe for new wonders of natural history, it was recognized that eastern North America and eastern Asia share many kinds of plants that are found nowhere else in the world. It was in 1846 that Harvard Professor Asa Gray made the first carefully detailed comparison between the floras of North America and, in this case, Japan. In the years since then it has become clear that only one other part of the world shares with our own southern Appalachians the great similarity to the old Tertiary forest. That is eastern Asia, most especially central China. This spot of earth, too, has been spared the changes that have affected the temperate forest everywhere else, and it, too, is a land of hills and valleys with an abundant year-round rainfall and a real but moderate winter.

The forests of central China broke into the news in 1944 when living specimens were found there of a tree that was well known as a fossil but thought to be long extinct. This is a tall, fast growing conebearer that sheds its needles in winter but is otherwise much like its relative the redwood. The fossil plant had long ago been named *Metasequoia,* and for convenience of the nonscientific public it was now given the common name of "dawn redwood."

Soon after this remarkable discovery a scientific expedition went into the remote Chinese hinterland to study the living fossil on its home ground. This is a high valley about a hundred and fifty miles east of Chungking, and although as the crow flies it is only fifty miles from the populous Yangtze Valley, so far is it removed from the beaten path that the expedition had to walk in from the nearest town on the Yangtze for a distance of a hundred and twenty miles.

There they found a true hidden valley, shut in by low mountains literally on all sides, as the Shui-hsa River that drains it escapes from the valley by an underground passage in the limestone bedrock. This no doubt explains why, in such a densely populated part of the world, this valley was first settled only about two centuries ago, within the lifetime of the grandfathers of some of the old people whom members of the expedition talked to. The first settlers had found the entire valley filled with dense forest; but since that time most of the hillsides have been cut over for timber and charcoal, and the flat valley floor is given over to rice paddies. Only in some of the side ravines that descend from the mountainsides and open out onto the valley floor does the *Metasequoia* appear to be growing under truly natural, undisturbed conditions. Many of the larger trees must have been well grown when the valley was settled, as counts of the growth rings made on borings from some of the trunks show they are about three hundred years old. One of the largest trees has a small, old temple at its base. The local people have always planted the tree around their farmsteads and along roadsides and riverbanks, and since they do not use the planted trees for any practical purpose, this seems to be an aesthetic matter.

Although *Metasequoia* grows vigorously when planted in a variety of places, both in China and elsewhere, it reproduces itself from seed only in the very special conditions of its native mountain ravines. There the seedbeds and natural nurseries are cool, dark, damp places under a dense tangle of shrubs and vines. The seedlings are very tolerant of the deep shade, but they grow slowly at first, and it takes them a while to push through the thickly matted overhead growth. Once they break through to the brightness above, however, the young trees grow rapidly, and it does not take many years for them to reach a height of a hundred feet.

The virgin remnants of *Metasequoia* forest are as strikingly like the temperate forest of Tertiary times as are the cove forests of the Smokies, and as richly diverse. One tract of a thousand square meters was studied intensively. On it were found twenty-seven different kinds of trees, with six more kinds growing nearby, and in the tangled underbrush there were fifty different species of shrubs and vines. Of all these, *Metasequoia* itself was most abundant, with many individuals of all ages and sizes. Next came the evergreen Cunninghamia or China fir, and the broadleaved chestnut and sweetgum.

No climatic data were available for that secluded place to provide a comparison with the climate of other forested regions; but the presence of rice fields shows that the growing season is long, warm, and wet, and the expedition learned that although winter rainfall is rather light, there is much high cloudiness and winter weather is generally damp. The surrounding mountains are high enough to shut out the wind, and severe cold waves are almost unknown. Here is a spot that has apparently been untouched by climatic or geographical changes for millions of years, and it is no wonder that a fragment of ancient forest lingers on here long after it has disappeared from other parts of the world.

Many of the plants that grow in both eastern Asia and eastern North America are so similar that experts give them identical scientific names. A homely example of this is the familiar skunk cabbage, *Symplocarpus foetidus.* Other pairs show detectable, though small, differences that have no doubt developed in the many generations since the two regions became so widely separated from each other. There are other, larger groupings of similar species. Though these are not identical, the relation is close; and in the modern world it is only in the Far East and in eastern North America that you will find, as familiar examples from a long list, tuliptree, sweet gum, catalpa, witch hazel, snowberry, spicebush, partridgeberry, phlox, and trillium.

Although it was the breaking of old land connections that first disrupted the ancient circumterrestrial forests, the floras and faunas of the separated regions were also strongly affected by the cooling climate. As conditions in their homelands changed, plants and animals that were able to adjust, or perhaps were already potentially adjusted, continued to live on in their old haunts. But most migrated south or, what is the climatic equivalent, from the cooler uplands down into milder low places. Totally new kinds of habitats also appeared, and into each of these moved the few things that could make a living there.

In good times and bad there is always a slow evolution of new types of plants and animals; but the long record shows that as a general rule, climates change faster than most creatures can change to fit them. Consequently, survival is usually a matter of either being versatile enough to live on under the new conditions or else migrating to a new home. Of course individual plants do not get up and move, and individual animals seem to stay close to the place where they were born and reared. But each new generation must find its own compatible living place, and this it takes where it finds it.

In both North America and eastern Asia the southward migrants moved freely over large expanses of the continent. But in Europe the way south was blocked by mountains that extended from the Pyrenees to the Carpathians. When in addition much of the land became covered with ice, many living things literally had no place to go, and so became extinct. There are many cases where only one species survived the Ice Age in Europe while a number of its close relatives live on in America. An example is the English holly, sole European member of the genus *Ilex,* which in America has ten still living species. These include not only what is familiarly known as American holly, but also inkberry, winterberry, gallberry, and yaupon.

Many kinds of both plants and animals have become totally extinct. Others persist in a range so small that their survival seems extremely precarious. The redwoods and giant sequoias of California, each in its own special niche; the ginkgo, which survives only because it has been cultivated from time immemorial around temples in the Far East; the *Metasequoia* that was long thought to be already extinct: all these were widespread in the worldwide Tertiary forests. Since all of them still grow vigorously when brought under cultivation, even in places far distant and very different from their native lands, it is hard to find a reason for their decline that is anything more than speculation. Perhaps the special nature of the *Metasequoia* seedling nurseries provides a clue to one major hazard in the process of self-perpetuation.

By the time the ice began to form far in the north, plants of the ancient temperate forest had already sorted themselves into the new and different habitats that had appeared. Some had spread down onto the damp flats of the coastal plain as it emerged from the edge of the sea. It is thought that these plants may have evolved far back in the history of the Appalachian region, at a time when the whole area was a low, sea-level plain. As the land slowly rose to its present height, small parts of the old surface remained as swampy flats perched on the newly elevated uplands. When the new coastal plain offered itself for colonization, plants from the vestigial flatlands, being already adapted to life in such a place, promptly moved down onto it.

Another group of plants had remained in the drying center of the continent as the rising western mountains cast a stronger and stronger rain shadow. These were the founders of the grassland vegetation of the prairies and plains. The oak-hickory forest of the Ozarks has descended directly from the Tertiary temperate forest; but during its long separation from the eastern forest it has been modified by a considerably drier climate. Presumably the far north had developed its own type of cold-resistant vegetation before the ice came down.

Only a corner of the rich Appalachian forest on its hill country extends into the Middle West. In this forest fully a dozen different kinds of trees are of major importance: beech, tuliptree (known to lumbermen as yellow poplar), several species of basswood, sugar maple, sweet or yellow buckeye, red and white oaks, and hemlock. Most especially characteristic are sweet buckeye and southern white basswood (not the basswood of northern woods). Still others are abundant although somewhat less universally present: both yellow and sweet birches, silverbell, black cherry, cucumber tree, white ash, and red maple. Among the smaller trees that do not reach to the top of the forest are dogwood, several kinds of magnolias, sourwood, striped maple, redbud, blue beech, hop hornbeam, holly, and serviceberry, the last known in the south as sarvis and in the northeast as shadblow. A profusion of small plants form many-textured carpets of flowers in the spring blooming season and a lush greenery of ferns in summer, with asters and goldenrods making another season of bloom as the summer ends. Beneath all this luxuriance the soil is mellow and porous, blackened to a depth of several feet with the accumulated humus of hundreds of thousands of years.

In the heart of this ancient Appalachian island, the forest is much the same in all but the wettest and driest places, although the relative abundance of the different trees varies from place to place. There beech grows even on dry hilltops and ridges, along with white oak and hickory, and only on the rockiest crests do the trees thin out to a rather open stand of pine. There is little in the way of floodplain to harbor a swamp forest, and often the streamsides have only a narrow border of moisture-loving trees such as the river birch.

Away from the center of this ancestral forest remnant, in southeastern Ohio and in southern Indiana and Illinois, living conditions are more varied and often more difficult in one way or another. In these less genial regions, many things may influence the exact composition of the forest: the direction in which a hill slope faces, the nature of the underlying bedrock or the soil derived from it, the wetness or dryness of the ground.

In the glaciated part of the country, where both soil and topography were drastically altered by the ice, conditions differ still more from the heart of the Appalachians. Since the ice has gone from the land, however, time and geological processes are slowly undoing its work, and the vegetation is responding in turn. Wherever there are indications of natural change, they point always toward an increasing richness and diversity.

Places that escaped the most recent glaciation have had 300,000 years free of ice. There, wherever the soil has accumulated a quota of earth-darkening humus, the forest is now almost as luxuriant as it is in the ancestral Appalachian homeland.

6. The Great Forest: Land of Maple

OF ALL the members of the ancestral forest, it is the sugar maple, *Acer saccharum,* that has taken over as master on the deglaciated lands. Over its wide range maple shares its kingdom to a certain extent with one or another of its ancient companions. In the milder part of its realm, its chief companion is beech. Toward the west, in the Driftless Area and beyond, beech yields second place to the northern basswood. On the north, around the upper Great Lakes, it is hemlock, white pine, and yellow birch that accompany the maples. But through all this large region, wherever the living is best, sugar maple strongly predominates.

Beneath the midsummer foliage of maples and beeches the shade is deep and dark. There is little or no undergrowth except where windthrow, disease, or some other local disturbance has made a gap in the forest canopy. The brighter light of an opening may favor the growth of a thicket of shrubs, along with seedling and sapling offspring of the various nearby trees. Hordes of young maples often form island-like mounds whose centers reach toward the brightness of the sky, their edges tapering off where the light grows dimmer.

Earlier in the year, before the tree leaves have expanded and shut out the light, there appears one of the delights of the maple forest — the large, dense carpets of spring flowers. As winter ends and the strengthening rays of the returning sun begin to warm the woodland floor, these small plants arise from the earth in a burst of rapid growth that soon unfurls both leaf and flower. Trillium or anemone, Dutchman's breeches, spring beauty, or blue phlox by the acre make a breathtaking sight, in its way more spectacular than the more varied display of the hill country forests of the southeast.

The blooming soon ends as one by one the plants set about forming their seeds and fruits. Before long they lay down buds with the rudiments of the next year's leaves and blossoms and store away a food reserve to fire the engines of another spring awakening.

All this teeming life quiets down by early summer, and the leaves of most of the "spring ephemerals" shrivel and dry. Soon only probing into the soil reveals the presence there of bulb or rhizome where the flowery carpets spread themselves in April and May.

In spite of the general fury of blooming and the prolific formation of seed that may follow, many of these small plants increase themselves chiefly by the spreading of stem and root. Although their aboveground parts rise from the soil as single shoots, the upright shoots are in most cases side branches from a creeping underground stem, or rhizome. Some of the plants grow in thick clumps that may become several yards across. Each clump, or clone, is actually a single plant that began from a single seed. Sometimes a slightly different color of flower or a distinctive pattern of lobing in the leaves indicates the oneness of a clone and sets it clearly apart from others of its kind.

When the rhizomes of such plants are carefully dug up and examined closely, they usually show scars of some sort where the upright shoots of earlier years were attached. This provides a convenient key to how fast the plants spread and a way of finding out how old a clone might be.

On the record of its rhizomes, one of the fastest growers

is the mayapple, which adds over five inches a year at the end of each underground branch. Solomon's seal adds slightly less than an inch, wood anemone about half an inch, while bloodroot grows at the modest rate of a quarter of an inch per annum. On the basis of such figures, the average age of mayapple clones in one piece of Wisconsin woods was found to be forty years. Clones of troutlily, which some call adderstongue or dogtooth violet, ranged from forty to more than three hundred years old, the average being 145! All this is a far cry from the apparently ephemeral nature of these delicate-looking plants.

As a clone grows very large and old, parts of it die and disappear for one reason or another, and it becomes impossible to do any more than guess at its probable age. After a long enough time the clones become so broken up that the plants are more or less uniformly scattered through the woods. When it reaches this stage, a forest has been little disturbed for a very long time.

By midsummer the floor of a maple forest is usually bare. By then the spring flowers have died down, and much of the last autumn's crop of fallen leaves has been eaten by earthworms and millipedes and the rest transformed in the slow fires of microbial decay. Although the tougher and tannin-rich foliage of beech and oak may remain over from one season to another, the soft leaves of maple, ash, and basswood are eagerly devoured by the lowly dwellers on the forest floor.

Maples and other soft-leaved trees have a generally beneficent effect on the soil they grow in, for the mineral matter that has been brought from the depths of the earth by far-probing rootlets to become part of the living treetops for a season is returned with the leaves to the ground. There it feeds a generation of microbes and insects and eventually works its way, several times transformed, into the earth again. The net effect of this "pumping," and in the long run cycling, is a gradual enrichment of the upper part of the soil. The enriched layer under the forests of the glaciated region is not very deep at this stage in its history, but given a few hundreds of thousands of years it will come to resemble the deeply dark soil of the ancestral Appalachian forest.

Like other places where sugar maple is abundant and autumn weather is clear and sharp, these woods have a brief glory every year when the maples put forth their gold and scarlet trumpet call. The golds of birch and an occasional hickory, the deeper reds of the oaks along with the contrasting somber tones of evergreens and the pale trunks of birch and aspen combine to make a brilliance of color and texture that even those who see the spectacle every year of their lives never come to take entirely for granted.

Of the animals that live in the maple forest, the most typical is the white-tailed deer, so much so that the entire assemblage of things that live there together is sometimes identified as the "Deer-Maple Biome." Although deer and foxes, opossums and skunks, raccoons and gray squirrels are most visible, the numerically greater population consists of rodents — chipmunks and mice in variety — and insectivores — shrews and moles. In a northern Ohio woodland that was studied in great detail, the number one mammal proved to be the tiny short-tailed shrew, followed by the woodland race of white-footed mouse. Birds in the forest are legion, with probably more vireos and wood peewees than anything else. But it is the still smaller creatures that live in rotting logs and leaf litter and in the soil itself that are overwhelmingly numerous.

Various attempts have been made to arrive at a reasonable estimate of the total animal population of such a forest. One biologist offers the following as average numbers of individuals for one hectare (about two and a half acres): 1 bird; 3 mammals; 3000 snails and slugs; 20,000 centipedes, millipedes, and sowbugs; 35,000 spiders and their kin; and 225,000 large insects. Even that does not exhaust the census, but it will help to put the deer and the birds in their proper places in the life of a forest!

Not much is left of either maple-beech or maple-basswood forest to show what it was like in its primeval state, for those regions have developed into some of the nation's most prosperous and intensely farmed land. The small bits of forest that remain in an even relatively undisturbed state are confined to places that are impossible to cultivate, such as steep ravines or spots more or less surrounded by cliffs. A few tracts have been spared because of someone's reluctance to despoil a beloved piece of woodland, and a number of remnants have been permanently preserved by more or less public action. There are tracts of quite impressive beech-maple woods in the Chagrin Reservation near Cleveland, in Warren's Woods near Three Oaks, Michigan, and in Turkey Run and Shades State Parks near Crawfordsville, Indiana. In Minnesota, good maple-basswood stands remain in Northfield Woods and beside Lake Minnetonka, near Minneapolis.

In a forested country, every farm originally had a woodlot set aside. Many of these survive, and one might expect them to provide samples of native forest. But though they are native enough, they are in a sadly depleted state. For many years the best trees have been taken out for use as lumber around the farm. All too many woodlots are used as pastures, or at least as pasture annexes where cattle can find shelter from the weather. Many a pastured woodlot has a browse line so sharp it might have been cut with a machine. Even where grazing is less severe, the soil becomes trampled and compacted, and any tree or shrub seedling that manages to get a start is soon demolished. Consequently, old woodlots often have no young growth at all and consist only of an open stand of large, decrepit trees with a cover of weedy grass beneath them.

Moreover, the total area of farm woodlots is small and it shows a steady decrease over the years. For example, in Shelby County, a completely agricultural area in western Ohio, as long ago as 1939 woods of all kinds occupied only 6.7 percent of the land; and in Green County, Wisconsin, on the Illinois line, the percentage in woodland fell from 4.8 in 1935 to 3.6 in 1950 and is probably even lower now.

Farther north, where beech and basswood give over to hemlock and yellow birch and where large tracts of pine alternate with the hardwoods, the soil is not very good for farming. There is a certain amount of dairying, but only in limited areas. Here the original forest was destroyed in a great sweep of lumbering. The time of clear-cutting came rather late in the upper Lakes States, not starting in some areas until the 1890's and lasting into the 1920's. In the earliest days only the magnificent great pines were cut. Some of these grew in large, practically pure stands; others were scattered singly or in small groups in the broadleaved forest. Even areas that today are considered virgin usually contain a few isolated stumps of what must have been enormous white pines. Some tracts have merely been "high-graded," where pines and the best of the hardwoods were removed without actually destroying the forest.

Despite the widespread activities of the loggers, there is far more old-growth forest left in this northerly country than in the predominantly farm country to the south, and much that was severely cut is now in some stage of regeneration. There are still some small tracts of virgin forest and others of at least relatively undisturbed old growth in the Porcupine Mountains and the Tahquamenon Wilderness in Michigan's Upper Peninsula and in the Flambeau River Forest and the Menominee Indian Reservation of northern Wisconsin.

The pine-hemlock-birch country has a distinctly different atmosphere from the regions of beech and basswood. This is readily apparent to anyone traveling north or south in Michigan, where the entire aspect of the country changes around Midland or Clare or Muskegon. To the south there are farms, and the country has a peopled look. What little woodland there is seems rather mild and tame. Toward the north the works of man thin out, the woods take on a wilder look, and a traveler is glad enough to put in at dusk to some settled place or to his own small home-place of a campsite.

In all the land where sugar maple is so strongly dominant, it requires no very sophisticated eye to see that these trees have preempted the best of the living places, where the soil is moist and yet well drained. On valley sides and morainal ridges, provided the soil is not too sandy and dry, the gentler slopes are covered with maples; and on wet till plains and extinct lake beds that seem absolutely flat, stands of maple may reveal the existence of faint mounds and ridges where drainage is slightly better. Toward the west, where the problem is more often a lack of moisture, the maples have taken over northern slopes where the sun shines less fiercely and deep valleys where the air is less affected by drying winds.

As the climate grows progressively drier, sugar maple becomes more and more sharply restricted to cooler and moister situations. It reaches its extreme western limit in some of the steep canyons of central Oklahoma. Beyond Indiana, however, over all of Illinois, Iowa, and southwestern Minnesota, sugar maples no longer dominate the scene but are outnumbered by oaks.

Sugar maple must be one of the most prolific trees in existence. It produces such prodigious quantities of seedling offspring that even the most technical scientific reports, full of graphs and tables of data, use words like "tremendous numbers" and "amazing" and "astonishing." In a good year an acre of forest may yield more than two million seeds, each one capable of producing a mature tree. There are, however, many hazards along the way between a sprouting seed and a forest giant.

This has been documented with actual counts and calculations. For example: In a southern Wisconsin woods the 1953 maple crop consisted of more than 2.5 million seeds per acre. Of these something over 1.5 million survived intact and uneaten to sprout the next spring. By that fall they were down to 80,400 and by the next spring to 45,640. In the third spring only 1 percent of the original seedlings remained; but that still added up to 14,320 little maples per acre.

Although maple seedlings may be able to endure even the heavy shade under their parent trees, any break in the forest canopy gives the infants a strong boost and greatly increases their chance of survival. In such an event it does not take very long for some of them to push up into the top of the forest. But the number that attain this success is still more sharply reduced. The loss of one tree with a crown 20 to 25 feet in diameter would make a bright spot below for some 15,000 new seedlings. These might be reduced to 150 in three years. By the time they are 6 feet tall, perhaps 40 or 50 remain, and of these only one will find a permanent place in the forest roof. At that time it might be 6 inches thick and 30 years old if it has grown vigorously all its life. On the other hand, if it has been overtopped and shaded by other trees, it might be only a half-inch-thick sapling at the age of forty.

Once they are solidly established, maples may grow to a large size and live to a great age. Trees 400 years old and 40 inches through the trunk are not especially rare, and ages of two centuries are quite common in old-growth forests.

The sugar maple probably derives its great advantage over its competitors from the combination of prolific production of seedlings, extreme tolerance of shade, and the ability to survive a long time in a much suppressed state and then go into a spurt of rapid and vigorous growth when conditions improve. Being able to reproduce in its own shade means that maple can grow in the same place generation after generation and never lose its dominant status in the forest.

It has long been observed that sugar maples growing in different parts of the country have slightly different types of leaves. On this basis some botanists separate out one type of black maple and give it its own name, *Acer nigrum*. The leaves of this form are somewhat hairy on the underside and are less deeply lobed than those of the sugar maple proper, *Acer saccharum*. But there are many intergradations between the two, and not all agree that they are really two different species.

There is a certain order in the situation, however, for a close look at the geographical distribution of the different leaf types shows that "pure sugar maple" is the characteristic form from the northern Great Lakes east to the St. Lawrence Valley and then south along the Appalachians, while "pure black maple" is found only in the Ozarks. In the Midwestern regions between, from Missouri to Wisconsin, the trees show mixed characteristics.

The explanation offers itself that the Ozark and Appalachian types became differentiated from each other after the ancestral Tertiary forest had been divided by the meeting of the ice sheet with the long arm of the Gulf that once filled the lower Mississippi Valley. Later, as the last of the ice retreated northward and disappeared, the Appalachian type migrated north and then west across the upper Lakes, skirting east of the always drier central part of the country. This migration may have gone on during the warm interval of several thousand years ago, when the midcontinent was even drier than now. In the cooler and moister times since then, the two populations have spread far enough to meet in the lower Middle West and mingle to give the situation we now find.

Second only to sugar maple in its ability to endure heavy shade is the beech. It is less tolerant of dry soil, however, which probably accounts for its absence from the western part of the realm of maple. It is not nearly so prolific with seeds as maple and only occasionally produces a good crop. Historical records show that from 1853 to 1893 there was an average of five years between heavy nut or mast crops. Moreover, some crops in modern times have consisted largely of empty shells, a strange kind of production indeed.

Unlike maple, beech has a strong tendency to form sprouts along its roots, which spread widely just beneath the soil surface. This happens even without any injury or disturbance, and most of the increase in beech seems to take place this way. Since the sprouts are attached to their large parent, they are less dependent on their own resources than small seedlings, and they stand a much better chance of surviving to grow into large trees.

The shallow root systems of beeches probably account at least in part for their ability to grow in wet soil, where air is available only near the surface. This would also explain their sensitivity to drought, since they cannot reach the moisture that may still be available at somewhat deeper levels. Beech roots make a dense, fibrous network in the top few inches of the soil and offer stiff competition to any other plants.

Where beech approaches its western limit, along the southwestern side of Lake Michigan, it extends up the stream valleys that drain into the lake. These small valleys, only ten or fifteen miles long, are quite steeply cut into the flat landscape, and the onshore breezes of summertime funnel into them. Apparently the cool, moist air coming from the lake makes just enough difference to allow the beech to grow. It disappears entirely in the vicinity of Lake Winnebago.

The foliage of beeches is very thick, and a year's crop adds up to a substantial bulk. The leaves do not enrich the soil like those of most other hardwoods, but like oak leaves they are acid in reaction and contain much tannin, so that neither insects nor worms nor the bacteria of decay consume them very rapidly. The net effect is that they remain matted on top of the soil and do no particular good for such other plants as are able to survive the competition from shade and roots.

West of the range of beech, northern basswood takes its place in the otherwise similar forest. Maple and basswood form the typical cover of the Driftless Area and reach in a diagonal band northwest across much of Minnesota. Around Minneapolis the strip widens into a sizable area. Here the original trees grew very large and dense for that part of the country, and the French explorers called the region the "Bois Fort" or "Bois Grand." The name survives in translation as the Big Woods.

Basswood is another profuse sprouter, but it sprouts chiefly after the old top has been cut or damaged by fire. The new sprouts appear in a ring, often at a little distance from the base of the parent tree. If those sprouts are damaged, new ones appear on the outer side of the ring. When eventually a generation of sprouts survives long enough to grow into large trees, they may stand in a perfect circle, like a giant fairy ring. Even when they do not form rings, large basswoods often appear in clumps. Seeds are produced fairly regularly, but they germinate poorly except in damp soil, and the seedlings need strong light to develop well. Consequently, where there is much sugar maple, basswood increases chiefly from sprouts.

Around the upper Great Lakes and eastward across Canada to the Maritime Provinces and New England, although beech and basswood are widely distributed, they have a relatively minor role in the forest. Instead, the conspicuous companions of sugar maple are hemlock, white pine, and yellow birch. More southern trees such as sweetgum, tuliptree, and buckeye are completely absent.

Hemlock is another extremely shade-tolerant tree. Although it needs more light for active growth than sugar maple, it too can survive for a long time in light so dim that it scarcely makes any new growth at all. Many small, straggling hemlocks with stems less than two inches thick have been found, by counting the growth rings, to be over a hundred years old. Stumps of large and flourishing trees often have a core that reveals the meagerness of their early years. Sometimes after a period of rapid growth a tree once more becomes shaded or crowded, and this too shows in a narrowing of its growth rings. With such uncertainties, it is almost impossible to estimate the age of a hemlock from its size, even more so than for most trees.

Hemlocks are not sprouters, and although they produce

quantities of good seeds, the seedlings require just the right conditions to become successfully established. In an undisturbed forest, the best places seem to be moss-covered logs and stumps of old hemlocks and pines. They can also grow on the sides of the tip-up mounds that are formed when an old tree falls over and its roots pull up a mass of soil as they are rent from the earth.

Small hemlocks may also appear in quantity on bare soil in exposed places such as the banks along road cuts or at the edges of ravines where the surface is swept bare by the wind, provided the soil remains moist enough. Since the seedling root is only about an inch long at the end of its first year, any drying of the soil can be rapidly lethal, and like the maples, hemlocks suffer a high infant mortality. The tiny root of the seedling also limits its ability to get through a mat of loose surface litter, which may account for the usual scarcity of seedlings beneath large hemlocks.

Yellow birch is the other chief companion of maples in the more northern part of their kingdom. Like the hemlock, it often begins life in a flock of seedlings growing on a mossy log. In time, these may develop into a long, straight row of trees, sometimes propped up on roots that originally straddled a fallen tree trunk that has long since disappeared. They may also start in large numbers on bare mineral soil if it is damp enough in spring. Since birch requires more light than hemlock for a good start in life, it needs some opening in the forest roof, although not necessarily a big one. If undisturbed, the trees may grow large and ancient, and a 300-year-old yellow birch may be 100 feet tall and 4 feet thick.

The advent of European man brought great changes to the realm of maple. When he came as a farmer, one of his first tasks was to improve the drainage of his land, for in its primeval state a staggeringly high proportion of the Middle West was waterlogged or even flooded for at least part of the year. It was not only low-lying flats that were constantly wet. Even in such a place as the rolling Appalachian Plateau of northeastern Ohio, about nine-tenths of Trumbull County had to be underlaid with drainage tiles before it could be farmed.

Widespread drainage often brought a general lowering of the water table. This affected not only cultivated fields but also the nearby woodland. As a result, many tracts of swamp forest were converted from elm, ash, and soft maple to pin oak or almost pure beech, and other places where beech had been strongly predominant saw a great increase in the amount of sugar maple.

Another of man's activities that has brought even more change in the forest is lumbering. Early operations often involved selective cutting, taking out only the more desirable large trees. These were usually oak, tulip, walnut, and black cherry, and, in the north country, pine. Such a procedure also produces a great increase in the abundance of beech and especially maple.

Even after clear-cutting, if no further disturbance takes place, a new forest grows up that is much like the old one, except for an increase in the numbers of light-demanding trees. In the north, white birch, aspen, and sometimes pin cherry commonly seed into any kind of clearing; but stump sprouts from maple, beech, and basswood form a dense, fast-growing coppice that soon overwhelms most small seedlings and restores the forest to its former composition.

The most drastic kind of disturbance in a forest is fire. A light burning may do no more than favor the fire-tolerant and encourage the sprouters; but even a moderate fire often opens the woods enough to increase the dryness of the forest floor and decimate or even eliminate many of the more delicate small plants that grow there. A severe fire may bring drastic changes that take a long time to undo. In addition to killing or even totally consuming the existing vegetation, it often destroys the humus mat in which many northern plants grow.

All the evidence shows that there have always been disasters in the forest. Fire scars buried deep in the wood of old trees and charcoal embedded in the peat of marshes and bogs date back centuries, to times before Europeans even knew that North America existed. Some of the fires were probably started by Indians, others by lightning. Other kinds of disasters also opened the forest to light. Early travelers and explorers were sometimes severely hindered in their goings about by wide swaths of wind-downed timber, and the old land survey records contain many notations of large "blowdowns." The Battle of Fallen Timbers was fought at such a place in the wilderness near Toledo, Ohio, in 1794. More often smaller openings were made by windthrow of small groups or even individual trees, or by the destruction of an isolated tree by lightning stroke or disease. It is such chance happenings that allow a variety of plants to persist in the forest as a whole.

Knowing the growth habits and requirements of the various trees makes it possible to estimate how long it has been since a tract of forest has been disturbed. Aspen and white birch are rather short-lived, and their presence shows that no more than a few decades have passed since there was a real opening. Black cherry, red and white oak, yellow birch, hemlock, and white pine also indicate a disturbance at some time in the past; but they are all capable of living for several centuries. In that length of time many of them succumb to one or another of the vicissitudes of life, and they gradually become intermixed with other species. The oldest survivors are often scattered so widely through the forest that no sign remains of their common origin or its cause, and they appear to be permanent and regular members of the community.

There are many places where the long past history of a forest can be deciphered in some detail. A piece of woodland in southern Wisconsin can serve as an example. There the oldest trees were found to be sugar maples, some of them dating back to the early 1700's. As might be expected, maples of all ages were found in abundance, with the usual tremendous numbers of young ones. There were also several other kinds of sizable trees, and when the growth rings were counted on small cores bored out of a large number of them, it was found that all the red oaks and many white oaks, butternuts, black walnuts, and still others dated to a period between 1830 and 1855. In those years something must have happened that opened the woods to the invasion of many other species into what had been a more or less pure stand of sugar maple. In view of the history of the area, what probably happened was a period of cutting for log cabins and then houses on the nearby prairie. After that the forest was apparently left undisturbed, for the only new trees added since then are scattered singles of slippery elm, basswood, and ironwood.

The same kind of detective work has been done for a number of hardwood stands in the Nicolet Forest of northern Wisconsin. Some of these were virgin forest, others had suffered minor logging but had never been clear-cut. In one of the virgin stands that were studied in detail, the oldest trees found were a white pine stump 350 years old and a sugar maple of 291 years, plus several large, mossy fallen trunks of sugar maple and yellow birch. Since this study was done in 1940 and no one knows when the large pine was cut, it must date back at least to the year 1590.

A number of large maples, yellow birches, white pines, and hemlocks clustered around 245 years old. These were scattered in the forest, and it was only by virtue of boring practically every large tree in the area that the existence of this age group was detected. They indicate that a fire in about 1695 killed enough of the then existing trees to allow a new generation to become established before the forest top closed in again.

There followed more than a hundred years of undisturbed growth, lasting until about 1800. There is no indication of any widespread disaster during that century, but a sizable fraction of trees date from between 1750 and 1800. This pattern suggests that, as so often happens, the fire of 1695 may have been followed by a generation of aspen. After about fifty years, the aspen would have gradually died out, since it could not replace itself in the shade of the other trees that had grown up through it. No trace of the aspen, if aspen it was, remains; but its existence is strongly suggested by the presence of a generation of other trees that were able to start in the brighter patches presumably left by the dying aspens.

In 1800 the forest was swept by another fire that left char marks on some of the tree trunks, now buried beneath newer wood. This allowed the start of yet another generation of light-demanding trees — white birch, yellow birch, and white pine, as well as some hemlocks, all still living at ages of about 140 in 1940. There is another tract near the study area that was also severely burned in 1800, but it had a different history and now bears an even-aged stand of large white pines.

When the ages of trees in many other tracts of long undisturbed northern hardwoods are examined in this way, it is found that in something like half the cases the trees are essentially all the same age, suggesting that they developed after some widespread devastation. Others are of thoroughly mixed ages, indicating a long period of growth with no disturbance greater than the occasional loss of a tree or two here and a small group there, just enough to maintain a variety of species and ages. Many other stands have trees that date chiefly from two or three separate and sharply limited time intervals.

Where nothing happens at all to upset the steady growth of trees, only the most shade-tolerant kinds can find a foothold on the dark floor under an unbroken forest canopy. There the young plants hold a slender tenure of life, eventually to fade away unless some brightening influence at last allows them to push rapidly up to take a place in the forest roof.

Since maple can out-tolerate even hemlock and beech, it is conceivable that if the forest were left literally and totally undisturbed for a few centuries, a large part of North America might well grow up to one immense and monotonous pure stand of sugar maples.

7. The Great Forest: Land of Pine

ALTHOUGH MAPLE takes over wherever the soil is loamy and moist, the drier sands belong to the pines. On the beds of extinct rivers and lakes, on the outwash set down from moraines and melting ice fronts, wherever the surface is sterile and coarse, there the pines come into their own.

All around the thumb and fingertips of Michigan's mitten and over the eastern half of the Upper Peninsula, where Lakes Saginaw and Algonquin and still others stood while the Great Lakes were evolving; on the sandy expanses washed out from the massive moraines of the north in Wisconsin and Minnesota; on the beds of once great floodways that carried off glacial meltwater: in all these places once stood the great pineries of the Lakes States.

It was white pine that grew to most memorable proportions. At the time they were cut, many of the magnificent trees of the virgin crop were about 400 years old and towered 200 feet into the air. Some of those recorded as bearing trees by surveyors for the Land Office had trunks of 7, 8, and even 10 feet in diameter. Such supergiants must have been older even than the main stands.

Although practically all of the virgin pine has long since gone down the lakes and rivers to build the farmhouses and the burgeoning cities of the prairie region and to make ties for the west-running railroads, a few tracts have survived to the present. Nothing at all remains of the most spectacularly large trees; but those were always sparsely scattered, even before there had been any cutting, and they were always the first to be taken.

One beautiful tract of virgin pine provides the core of Hartwick Pines State Park, in the northern part of lower Michigan. Here is a tract of eighty-five acres that somehow escaped the loggers. It stands on a sandy ridge and extends down onto the edge of a flat sand plain. Most of the large trees here are white pines, although in some places there are red pines mixed in with them. The largest tree of all is set apart like a monument, with a little fence and a plaque, and has its own name: "The Monarch." It rises to a straight and majestic height of 155 feet, with more than 70 feet of clear, smooth trunk to the lowest limb. At breast height it is four and a quarter feet thick.

Throughout most of the Hartwick Pines there is a second layer of trees whose tops rise to about a hundred feet. These are hemlocks and such northern hardwoods as beech, oak, yellow birch, and the ever abundant sugar maple. The undergrowth is rather sparse except under slight breaks in the overhead canopy. There where the light is a little less dim the ground is covered with a virtual carpet of maple seedlings. Up on the tops and sides of mossy stumps and fallen logs there are seedlings of yellow birch and hemlock. Small pines, however, are all but nonexistent. This lovely forest is probably a fair sample of the primeval white pine lands.

Farther west in Minnesota, where the climate is somewhat drier, red pine becomes more abundant, although both red and white are common throughout the Lakes States. In Itasca State Park, 200 miles north and somewhat west of Minneapolis, one of the special charms is the presence of huge red pines. Many of these date from the years between 1715 and 1735, and there are several stands dating from the early 1800's. Here, too, the trees are tall and straight,

Cabin of large logs from the virgin pine forest of the great days of lumbering. Itasca State Park, Minnesota.

with long, clean boles rising to the lowest branches. The sunshine slanting through a grove of them on a summer evening, striking the bright red bark of the upper trunks, makes it abundantly clear why they are called red pine, although in the Middle West the name Norway pine is commonly used. This park, too, is a lovely place. It lies on a tract of irregular moraine and offers not only the sandy reaches where the finest pines grow, but lakes and bogs and bits of hardwoods, as well as the remote north-woods source of the Mississippi River.

Jack pine never grows to the proportions of either its white or its red brother, and it always has a rather scraggly look. No reserves have been set aside in its special honor. But on the driest, most sterile sands, little can match the hardiness and persistence of this tough and enduring tree.

More than any other plants, pines are the offspring of fire and sand. Only by fire is the land cleared of both vegetation and loose surface litter, so that the full light of day falls upon bare soil. Windstorms may blow down wide swaths and make openings in the forest; but drying foliage, trunks, and boughs of the downed trees, or vigorous new sprouts from roots and stumps shade the soil surface, giving a strong advantage to the shade-tolerant hardwoods.

Once the earth is exposed, it becomes seeded over with whatever the surroundings have to offer. In a good seed year, pines are very prolific; and although the seeds are relatively large and heavy, their papery wings act as effective little sails in a stiff breeze, and they flip and flutter for some distance from the parent trees before they come to rest. The smaller, lighter seeds of birch and aspen, also sail-equipped, can blow for long distances, and in the high winds of winter and early spring they skitter and skate far over the slick surface of crusted snow. Other seeds that are borne in pulpy fruits or berries may be widely distributed by birds. Pin cherry is a good example of this.

Which of the potential pioneers actually take over a given burned area depends on a number of things, including a certain element of chance. The light seeds of aspen and birch have the advantage of great numbers and high mobility, and on distinctly heavier and moister soils they grow rapidly from an early age, commonly outstripping any seedling pines. In such places the first trees to establish themselves are likely to be white birch and trembling aspen, often with a certain amount of pin cherry, especially if the soil is clayey. These soon provide shelter for more shade-tolerant invaders, and though the ultimate forest that develops on good soil may have a scattering of white or red pine, it consists predominantly of broadleaved trees.

If there are many pines in the vicinity, they may seed into a bared place so thickly that they crowd out a lesser population of hardwood seedlings. White pine can do this on soil that is moderately fertile and moist. Somewhat poorer sites tend to favor red pine, and the most hopelessly barren sands remain for the rugged jacks.

Jack pine is above all others a fire tree. Only a few of its cones open and release their seeds until they are exposed to heat that is intense enough to scorch or even consume the tree that bore them. Until this happens, most of the cones hang tightly closed on the trunks and branches. Sometimes they persist so long that they become buried by the thickening wood to which they are attached.

A forest of pure jack pine is usually dry enough to be highly susceptible to fire, and when the almost inevitable happens, the cones open in the heat and release their seeds to start another generation. A hot fire may also burn off the organic litter, leaving a sterile sand surface that can support little else than jack pine and its typical undergrowth, and even that may be sparse if the sand is very poor. Since these trees can produce cones and seeds at as tender an age as five years, fires recurring at moderately short intervals can perpetuate jack pines on their sand plains indefinitely.

The rugged character of jack pines extends beyond their tolerance for dry, sterile soil and their way of exploiting fire. They seem to be indifferent to extremes of temperature that would do in almost any other trees. In open spaces in plantations, the soil surface has been known to reach 175° F. and to stay above 130° for eight hours at a stretch. On the other hand, in the same places frost may occur on almost any night of the growing season.

Seedling jack pines grow slowly at first; but after four or five years they seem to spurt and develop vigorously into thrifty small trees. As the ground beneath them becomes sheltered a little from the most extreme heat and dryness, seedlings of other pines and sometimes a few oaks may appear among them.

Beneath the developing evergreens the annual shed of needles slowly accumulates on the sandy surface. Their strongly acid nature makes them resistant to decay, and they soon form a loose mat, pleasantly springy under foot, that may become several inches thick. Water percolating through the mat becomes acidified, and as it works its way down into the soil below, it dissolves out much of the more soluble mineral matter, especially the limy components.

Gradually the soil takes on a layered pattern. Beneath the organic mat there is a very thin dark layer composed of a mixture of the original soil with what little humus develops from the damp underpart of the needle accumulation. Below this lies a strikingly ashy white layer, a few inches to a foot or more deep. This is sterile and acid and consists of almost pure sand. Its ashen appearance is the basis for the technical name of "podsol" given to it by Russian scientists. Below the white layer the subsoil is yellowish, and a foot or so down it grades off into glacial drift or outwash that is virtually unchanged from the state in which it was laid down during the waning of the Ice Age.

Not every kind of plant can grow in such a soil, even if its seedling root succeeds in penetrating through the thick and droughty surface mat. Many of those that can are members of the heath family, which are generally known as acid lovers. In these woods there are none of the rhododendrons or mountain laurel of acid soils in the east, but an abundance of low, shrubby blueberries and huckleberries, with patches of creeping wintergreen, pipsissewa, and glossy dark bearberry. Bracken fern and bunchberry may grow in dense patches; and where the shade is very thin there may be a quantity of aromatic sweetfern and a scattering of aspen.

As the young trees grow and begin to take on the aspect of a forest, their various growth habits begin to influence the course of events. For perhaps fifty or sixty years jack pines and aspens grow most rapidly and vigorously. Then they slow down and begin to succumb to the assorted vicissitudes that beset their kinds. In such places aspens commonly become infected with heart rot and a canker disease that make them especially susceptible to wind breakage, and in another ten or fifteen years little is left of them. The jack pines go somewhat more slowly, and although most of them are gone in a hundred years, a few may survive twenty to forty years longer, rarely more.

At about the time when the pioneer generation is beginning to break up, the red pines reach their maximum growth rate, and between 60 and 100 years from the first seeding-in they are the most vigorous part of the forest. By 130 years the red pines are rising above the tops of the remaining jacks, often attaining heights of 120 feet and a thickness of 30 inches or more. These are much longer-lived trees and they may dominate the situation for as long as three centuries.

By this time the forest floor has changed, and under the pines grow little evergreen mats of twinflower, shinleaf, and the persistent pipsissewa, lifting up in their seasons their elegantly crisp small flowers. Here live the trailing arbutus that scents the early springtime air, wild columbine, and the small polygala that is sometimes called gaywings.

The treetops now are lively with Canada jays, many kinds of warblers, and the chatty red squirrels that make a good thing of the plentiful pine seeds. Here the walker in the woods may start up a ruffed grouse or come upon a lumbering, bark-eating porcupine.

Meanwhile the white pines are also developing. At about 100 years they begin to pass the other pines. They are potentially the tallest of all, commonly growing at least 140 feet tall and 4 feet through the trunk. Since they are quite capable of living for 500 years, the forest for a long time may be essentially a vast expanse of large white pine.

All the pines originate early in the development of the forest, for as soon as the canopy closes over, pine seedlings are unable to survive in the deepening shade below. For a long time there may be no small trees of any kind; but slowly a few broadleaved seedlings begin to infiltrate where the light is a little brighter under a thin spot in the roof. Often the first invaders are red oaks and red maples.

Over the long centuries the pine-needle mat yields a certain amount of humus to the soil beneath it, and as the first hardwoods bring further changes, improving the soil's texture and fertility, an occasional seedling of the more demanding hemlock, basswood, yellow birch, or sugar maple succeeds in getting a roothold. By the time the white pines begin to die off in any number there is usually a well-formed understory ready to take over their places. The Hartwick Pines at the present time are approaching this state. Sooner or later the forest becomes heavily dominated by sugar maples. By the time it has become a stable and self-perpetuating community, a thousand years may have elapsed from the time of the initial burn-over.

This is the story as it occurs on sandy soil and in the absence of major disturbances. It has been reconstructed in many places and by many people, and it is surprisingly consistent from place to place.

Judging from the forests themselves, the great pineries of the upper Middle West must have started on just such a course of development sometime in the mid-1400's. The pine forest that preceded the present hemlock forests of western Pennsylvania also originated about that time. There are Indian legends of great fires about then, and archaeological evidence suggests that there were changes in the Indians' way of living around that time. While we have no way of knowing many of the details, many facts point tantalizingly in the same direction.

At any stage in the history of the forest the whole progression may be set back by another fire. In general, the closer a forest is to being a pure stand of sugar maple, the less susceptible it is to burning. Fires in dry slash or downed timber often die out as they work into the edge of a well-grown stand of hardwoods. Even when a fire does make headway, unless it is extremely hot, it does little damage to the underground parts of most broadleaved trees, and many of them sprout promptly and vigorously. Before long the forest may grow up to be very much as it was before, except for the loss of white pine or hemlock and probably the addition of a few less shade-tolerant trees such as yellow birch, red maple, or red oak.

On the other hand, severe fire in a forest of red or white pine may bring immediate and conspicuous changes.

There is usually at least a scattering of young hardwoods, and sprouts from these grow rapidly when there are no more pines to shade them. Consequently, a destructive burning may convert a pine forest more or less directly and dramatically to pure hardwoods.

The worst conflagrations, such as those that occur in times of severe drought, may consume not only the trees but also the surface litter and even the humus in the upper part of the soil. This reduces the soil's fertility so much that only the less demanding evergreens can grow, and the cycle must start over again from the beginning. Erosion of the denuded soil is another hazard, but in the generally level land of the Middle West this is less of a problem than it is in hillier parts of the country.

The influence of recurrent fire shows clearly in the history of Itasca. There are written records going back as far as 1803, and more of them since the expedition of Henry Rowe Schoolcraft in 1832, which established Lake Itasca as the true source of the Mississippi — the "verITAS CAput" from which he invented its name. Schoolcraft Island, lying in a fork of the lake, was described in 1836 as being "covered by a full growth of northern trees." In 1871, however, it was reportedly covered with underbrush; and in 1891, when the state legislature decreed that the source of the Father of Waters be set aside in a state park, the island bore one stately white pine, an occasional bur oak, and a dense growth of birch, basswood, aspen, fir, and diamond willow.

Many studies have been made of one aspect and another of the flora and fauna of the Itasca region, since the University of Minnesota has a forestry and biology research station in the park that provides a convenient base of operations. In the report of one of these studies, Stephen Spurr observes, "The single outstanding impression is that most of the park was burned over periodically during the nineteenth century." His paper has a photograph of a cross section of a tree eighteen inches thick and 227 years in the making. Its growth rings show a series of fire scars that date from 1772, 1803, 1811, 1820, 1865, and 1886. Other trees show that the fire of 1865 must have burned over the entire park area. The underbrush growing on Schoolcraft Island in 1871 would have been the aftermath of that fire, which destroyed the "northern trees" that were there in 1836.

For a long time after the virgin pines had all been cut, many areas were swept by wildfires practically every year. In some places this went on for fifty years or more. As a result, large tracts were reduced to what are commonly called pine barrens. These were really a sort of savanna, with a few scrubby trees scattered over an open expanse of grass and shrubs. On a hot summer day these desolated places had a bleak kind of charm when the aroma of sweet-fern blended with the fragrance of sun-warmed pines; and late in the season the dense stands of blueberry and huckleberry bore prodigal crops of fruit that hundreds of people could have simply for the taking. All this belongs in the past tense now, for since the 1930's, when the widespread and repeated burning was almost entirely stopped, the barrens have grown up to dense stands of jack pine.

Where the virgin forest consisted entirely of white and red pine, with no jack pine at all, the timber harvest was so complete that there were not even enough survivors to serve as seed trees. The occasional tracts that did become reseeded were usually burned over long before the young pines were old enough to produce seeds of their own. As a result, pine became practically extinct over large areas, and many miles of country were reduced to a continuous stand of stump-strewn weedy grass that was locally known as "stump prairie."

Even when fires were stopped, much of this remained in grass, partly because of the lack of a seed source, partly because a tight cover of grass and sedge is strongly resistant to invasion by other plants. For one thing, there is the mechanical problem for a seedling root to work its way down through the sod into the soil below. More important, the fine roots of a close turf are very effective in absorbing moisture from the soil. Even soon after a rain, the soil under grass may contain less than half as much moisture as soil in a nearby forest. The only way that such places can be converted to forest in a reasonably short time is by planting young trees that have large enough roots to reach below the grass roots.

It is now more than forty years since the uncontrolled fires were stopped. In that length of time, while the much-abused vegetation has been going about its business undisturbed, many changes have set in. For one thing, the millions of seedling trees that were planted on the cutover lands during the 1930's, most of them by CCC boys, have grown to usable size. This accounts for a very small area, however. A few other attempts at reforestation had been made before that, but they were very sporadic. The first effort seems to have been made in 1876, when some white pines were set out on a farm near Hancock, Wisconsin. Tree planting was also started on a substantial scale in the national forests in Michigan and Minnesota about 1910. But all this added up to very little, and most of the land by far has been left in the hands of nature.

By 1940 much of the burned-over region had become seeded in with aspen and birch. A survey made in 1950 found that about 40 percent of the commercial forest land in the Lakes States was then covered with a mixture of these two species, with the aspen predominating.

Although aspen will grow almost anywhere, on fertile soil that is both moist and well drained it develops at a spectacular rate. On such soil it can grow three feet a year in its first decade and eventually reach a height of ninety feet or more and a trunk diameter of nearly two feet. In a good situation, even after sixty or seventy years it may still be reasonably free of disease and decay.

This kind of productivity has generated considerable interest in the use of aspen as a crop tree. Ways have been found to use it for pulpwood, as well as for the traditional boxes and crates, and some landowners are practicing a kind of silviculture geared to this market.

In the absence of fire, aspen is ordinarily a one-generation tree. By the time it is what foresters consider mature, it usually has a well established understory of smaller trees of other kinds. If these are of no great value, which is often the case, the entire woods can be clear-cut and another generation of aspen allowed to develop from root sprouts. Sometimes the soil is disked to injure the more superficial roots and stimulate sprouting so that the new shoots will be dense enough and grow fast enough to keep ahead of any other trees that come in as seedlings.

Just what will eventually happen to all the land now covered with aspen is not entirely clear. Some of it will certainly be replaced directly with useful northern hardwoods — sugar maple, basswood, and yellow birch. Another fraction will no doubt go to spruce and fir, and a

After the pines were cut, a pioneer stand of aspen often followed. A new generation of evergreens is now growing up through these large aspens. Porcupine Mountains State Park, Michigan.

negligibly small amount will develop reasonably good pine. The rest appears headed for low-value hardwoods, such as scrub oak, red maple, ash, and elm.

For many years it was hoped that the cutover land would eventually grow up to pine again, but this clearly is not going to happen. If the pine is to return, it will have to be planted and protected during its youth from encroachment by the more rampant hardwoods.

In the bits of old pineland that have been set aside as museum pieces, such as the Hartwick Pines and Itasca, questions will arise about what to do when in the natural course of events the pines begin to go and are succeeded by hardwoods. Nobody is likely to countenance the kind of uncontrolled wildfire that originally gave birth to the pineries; but perhaps we will yet learn how to use controlled burning effectively to maintain the pines on their ancient sandy homelands.

With the ending of large-scale fires the future has become problematical for other living things, most urgently for the small bundle of animated fluff known as Kirtland's warbler. This little bird nests only in the young jack pines of the Au Sable River drainage basin somewhat north of Saginaw, Michigan. Its entire nesting range is encompassed within an area about eighty-five miles wide and a hundred miles long. Somehow in the course of its evolutionary history it took for its own the jack pine barrens where few other creatures live when they can find anything better. The only kind of place where Kirtland's warbler chooses to nest is a stand of jack pines between six and twenty feet tall, covering at least eighty acres, with plenty of openings among the clumps of trees and a thick ground cover of such fire-resistant plants as grasses, blueberries, bearberry, and brambles.

The critical factor in this situation seems to be a dense cover of living pine branches that lie close to the ground. When the trees are too small, their branches do not meet to give enough cover, and when they are too large the lower branches become shaded and die. Where they are just

Repeated fires on sandy soil give rise to jack pine barrens. Places like this, near Grayling, Michigan, are the home of Kirtland's warbler.

right the bird builds its nest in a slight hollow in the sandy ground beneath them, concealed under a small canopy of the ground-hugging lower vegetation.

In the days when fire swept freely and often over the dry sandy pine barrens of this part of Michigan, there were always many tracts of jack pine in just the right stage for Kirtland's warblers. By the time one stand grew too big, another had grown up from fire-scattered seeds to substitute for it. But with the limitation of fires, things began to change.

Ever since it was first identified in 1851, this bird has been known as a rare species. From the ornithological records, it appears to have been most numerous in the 1880's and 1890's, the years when lumbering was most active and fires most frequent in lower Michigan. During the nesting season of 1951 a group of thirty-some bird watchers attempted to count all of these birds in existence. On the basis of their record of 432 positively identified singing males, it is estimated that the world's total population of Kirtland's warblers is about one thousand — a total of perhaps thirty pounds of bird!

What will happen to this little creature in the future is still very uncertain, but an attempt is being made to save it from extinction. For this purpose several tracts of jack pine within its nesting range have been set aside for its very special benefit. These are on various state and privately-owned lands and in the Huron National Forest, near Mio, Michigan. They are to be managed on a sixty-year rotation cycle that will involve harvesting the large pines for lumber, then burning the remaining slash on the cutover area, all under carefully controlled conditions. The fire makes the old cones release their seeds and also reduces the competing vegetation. The seedling pines then grow vigorously, and about six or eight years after the fire they are the right size for the birds to begin to nest there. After another dozen years or so the trees are too big and the birds move to another place. The pines on each tract will then be left to grow until they are big enough to be worth cutting, and the cut area will then be burned again to start off a new generation of trees. Assuming the bird survives the heavy predations of cowbirds and that nothing untoward happens to it in its winter home in the Bahamas, this may be enough to save one small creature from going out of existence.

8. The Boreal Forest: Land of Spruce

As the great Pleistocene glacier retreated for the last time across the country, close behind it there followed a dense forest of spruce and fir. Since then the ice has outstripped the forest, and between them a zone of open tundra now reaches far into the Arctic; but as far north as there are any trees, the boreal forest of today is much like the one that first moved onto the ice-freed land.

This is a world-encircling forest, for from Alaska across Canada to Newfoundland and from Scandinavia to the taiga of Russia and Siberia, the same darkly tapering evergreen forms of spruce dominate the landscape.

It is only in northern Minnesota and Wisconsin that the upland spruce forest reaches into our Middle West. There it extends almost continuously from the wild country of western Ontario as far as Itasca and the north shore of Lake Superior. It covers Isle Royale in the northwestern part of the lake, and there are smaller tracts of it along the Wisconsin shore, on the Keweenaw Peninsula, and around the shore of Mackinac Island. The Door Peninsula between Lake Michigan and Green Bay has pockets of spruce-fir forest, and one of them gives a special charm to the vicinity of Bailey's Harbor there.

Except for the inland parts of Minnesota, all these places

USDA Photo by Freeman Heim

Much wild land remains in the north. The Boundary Waters Canoe Area in northern Minnesota has almost as much water as land. The glacier-scoured rock of the Canadian Shield has only a thin and patchy cover of soil, but it supports a beautiful boreal forest.

are within a dozen miles or so of the shore of one of the Great Lakes. These large bodies of water have a marked influence on the climate near them, especially where the winds are prevailingly onshore, for wind off the lake is never so hot in summer nor so cold in winter as the inland air, and it is also appreciably moister. When the temperature at Ephraim, on the west shore of Green Bay, rises into the nineties on a summer afternoon, it is likely to be a pleasant 70° at Bailey's Harbor just nine miles across the water. Moisture from the lake brings a somewhat higher rainfall in summer and a very heavy snowfall in winter, and four or five feet of snow on the ground is no novelty at all in Door County. Because of the tempering influence of the lake, autumn frosts are somewhat delayed, the beginning of spring even more so, and the growing season is correspondingly shortened.

In these cold, damp places, white pine and its common hardwood companions drop out of the picture. White birch and aspen range far to the north, as do jack pine and to some extent red pine; but these trees ordinarily come in after some disturbance, persisting only where a difficult soil suppresses their competitors.

In the best of the far northern forests, the dominant trees are white spruce and balsam fir. White spruce takes its name from the pale color of the waxy bloom on its stiffly pointed needles. The dark, glossy needles of fir, blunt-tipped and softer in texture, arrange themselves in flat sprays that give a characteristic horizontal emphasis to twig and bough.* The narrowly tapering treetops that are the mark of a northern skyline seem to have a practical function, for when the short boughs droop low under a heavy load of snow, a very small disturbance is enough to dislodge the snow and send it sliding harmlessly off to the ground.

Given time, spruce can grow over a hundred feet tall and develop a trunk two feet thick. Fir never seems to get so large, probably because it rarely lives long enough. Both trees are shallowly rooted and easily blown over, but all the hazards of breakage and disease seem to take a heavier toll of the firs, although they are extremely prolific and vigorous in their young stages. Both species have the resinous character that gives them a fresh north-woods aroma; but it is balsam fir whose fragrance intensifies as it dries, giving it a special charm as an indoor Christmas tree and making it a favorite for stuffing souvenir pillows, or for tucking into a coat pocket or a corner of one's luggage on a north country outing.

Besides the widespread and beloved white birch, perhaps

* Fir cones stand upright, spruce cones hang down. It would be much more convenient if one could just remember "spruce up," but such is not the case!

the most characteristic broadleaved tree of the northern forest is the mountain ash, close cousin of the European rowan. It is not especially abundant, but in late summer its clusters of bright red berrylike fruits are conspicuous where they light up a rocky ledge or reach out over a northern lake shore.

Where the evergreens grow thick, there is no undergrowth at all. Thin spots or small openings are decked with such northerners as clintonia, goldthread, or starflower. Bunchberry also grows far into the north. The wide ranging Canada mayflower, known to some as wild lily-of-the-valley, seems to be everywhere: it was found in every one of more than a hundred stands of boreal forest that Curtis and his students examined in Wisconsin.

In this moist forest many fungi, mosses, and lichens grow in the closely packed leaf mold that forms the ground surface. Lichens in variety often plaster thickly over rocks, fallen logs, and tree trunks and may hang in untidy tangles from branches and twigs.

The chief large animal here is the moose, a huge beast that gives its name to the "Moose-Spruce Biome." In summer one comes upon them standing shoulder deep in quiet ponds and streams as they feed on the underwater plants, at intervals lifting their great, dignified heads to stare, while the water streams from the broad blades of their spreading antlers, sometimes rakishly entangled with waterlily leaves. In other seasons they browse on the shoots of broadleaved trees and shrubs. The nipped-off twigs usually sprout behind the cut ends, and a well frequented moose pasture forms a characteristic thicket of stubbily branched stems leveled off evenly at about the shoulder height of a man.

Up to the early 1800's the woodland caribou also ranged into northern Wisconsin. Now the more common animal is the deer, and only in remote regions are there places wild enough for what remains of wolf, lynx, and wolverine. But such birds as spruce grouse, crossbill, and great horned owl still identify the land as the north woods.

Like other forests, this one is subject to destruction by fire or windstorm, as well as its own special pest in the form of the spruce budworm. Here, too, the pioneer trees of a new generation are chiefly aspen and birch, or sometimes jack pine. Here they are commonly followed by an even-aged generation of spruce and fir; and a familiar sight is a thinning stand of tall, pale stems with fluttering, light-textured foliage, and growing up through it a dark layer of narrow-topped conifers.

Like other lands not long released from the glacier's embrace, the boreal forest region is poorly and erratically drained. All through it there are vast areas covered with standing water and sodden masses of sphagnum moss interspersed with varying proportions of heath shrubs, black spruce, and tamarack. The Indian name for such places is muskeg, although the tamer-sounding European word "bog" is also commonly used.

Enormous expanses of muskeg still remain in the beds of Glacial Lake Agassiz and Glacial Lake St. Louis northwest of Duluth, as well as lesser but still considerable tracts in such places as the valley of the Tahquamenon River in upper Michigan. The large northern muskegs are better known from the air than from the ground, for except in winter when everything is frozen hard they are virtually impassable.

The Lake Agassiz muskeg is not confined to low places but spreads in an oozy blanket for miles up and down over the ever-so-gently undulating contours of the land. The faint and slowly moving drainage ways or watertracks can be traced by the zones of different kinds of vegetation that parallel their irregular courses. Such expanses of "patterned bog" are found all around the globe in the far north, and the Agassiz peatlands are the southernmost outliers of a truly boreal type of vegetation.

There are lesser tracts of muskeg all through the upper Lakes States. Still smaller areas of similar bog vegetation can be found all over the region covered by the latest ice sheet, as well as in a few hollows in the outwash that lies just beyond the glacial border.

In the north the black spruce of a well grown bog is not conspicuously different from the upland spruce forest nearby; but toward the south it contrasts more and more sharply with the surrounding vegetation. In the rolling Iowa prairie or the oak forest of northern Illinois the juxtaposition is striking. This no doubt explains why bogs long ago attracted the attention of students of the newly developing science of ecology, especially at the University of Chicago.

The more southern bogs commonly lie in undrained hollows, where the water is typically stagnant, often deep, and usually cold. The air, too, is stagnant and cold; for on still, clear nights when the lower air is chilled by contact with the rapidly cooling land, the heavier cold air settles to the ground. Then it begins to flow slowly downward, down the hill slopes, down the valleys, and when finally there is no outlet for it, collects in a pool at the bottom of just such a

A boggy stream near Shingleton, in the Upper Peninsula of Michigan. The sluggish water is bordered by floating mats of vegetation that may become thick and dense enough to support a stand of black spruce and tamarack.

low place as often contains a deep, undrained pond. In such a place the local climate is very different from that of the immediate surroundings; and while the milder uplands are covered with a more southern vegetation, down in the cold, wet hollow the plants of the boreal muskeg find themselves quite at home.

A bog is a cold place at any time. At noon of a summer day when the air nearby may be 80° F., down in the bog hollow it is likely to be in the sixties, while in the moss below, where plant roots grow, it may be a chilly 45°. Farther down it is even colder. Ice has been found only a foot below the surface of bog peat as late as July; and in spring many a bog plant has its top in full bloom or full leaf when its roots are still solidly frozen. The entire growing season is usually brief in such places. In fact, in bogs where the matter has been investigated, frost has been known to occur on almost any night of the year. Under such living conditions, one might well expect the vegetation to have a distinctly northern aspect.

The formation of a bog typically begins around the edge of a body of open water and works gradually in toward the center. Consequently, the plants become arranged in a series of more or less concentric zones, and the developmental stages can be read off in reverse order from the water's edge outward to solid land.

Fringing the water there is a thin, loosely woven network of floating plants, usually sedges. A little back from the edge the meshes become filled with sphagnum moss, and wherever this can support them, other plants appear on it. Even a thin tangle of moss may be traced over with the long, slender runners of cranberry. Where the floating mat becomes more substantial, it is dotted with heath shrubs such as leatherleaf, bog rosemary, and Labrador tea. Scattered among these are grass pinks, rose pogonias, ladyslippers, and others of the orchid tribe; and insect-catching pitcher plants and sundews add an exotic note. Outward the shrubs become thicker, and then trees begin to appear, tamarack first, then black spruce.

All these plants form a dense, spongy cushion that floats on top of the water. As the tangle of roots and stems and upward-growing sphagnum thickens, it gradually sinks of its own weight and eventually becomes grounded on the lake bottom, which at the same time is slowly building up to meet it. The free-floating part of the mat readily yields under the weight of an intruder, and even carefully placed footsteps set off a wavelike quaking motion that may carry for some distance. Where the cushiony mat rests solidly on the bottom it is still resilient enough so that jumping up and down will set even sizable trees to wobbling back and forth at some distance.

At the edge of the bog, just where the old lake bottom rises to the water's surface, there may be a narrow, moat-like strip of open water. This often has an outer edging of alder or red osier dogwood, and it is the common habitat of black ash, almost the only broadleaved tree that grows in a northern bog. It is also a favorite growing place for the tall, rangy poison sumac, and as John Curtis has remarked, "Actually, many of the better southern [Wisconsin] bogs owe their botanical excellence to this protective barrier of poisonous shrubs, since even their owners are hesitant about passing through the pale."

Bogs are found in every stage of development. Some have become practically obliterated and are hard to identify. Seen from the air, a rounded patch of vegetation that is just perceptibly different from its surroundings may disclose the existence of a former bog. Or digging into a damp place in the soil may reveal a compacted peat that includes much sedge and sphagnum moss. On the other hand, a small and poorly developed bog may consist of little more than a patch of heath-covered sphagnum growing at the edge of an open lake or in a stagnant cul-de-sac along a slowly moving north-woods stream.

As time goes on the thin margin of the floating mat slowly extends over the water, rarely adding as much as a foot in a year. Even this slow advance can be undone by various forces. If the mat reaches the edge of a moving current, it is likely to be pruned back at times of high water, when the current moves more swiftly. At any time high winds can rip away parts of it. A detached piece of bog mat moves around the water as the wind shifts, and if it bears shrubs or well grown trees, such a floating island can stir amazement if not consternation in its beholders.

An open bog changes slowly and may persist for a long time with only shrubs and other low plants. When eventually tree seedlings begin to appear, the first ones are generally tamaracks. These need the full light of the sun to do well, but given the right conditions they grow fairly fast. Since their roots rarely go more than a foot and a half deep, they can easily grow on the floating part of the mat.

Seedling tamaracks grow more or less neck and neck with the sphagnum in which they are rooted, forming new roots up the stem as the moss rises around them. Without this capacity, they would not long survive on the bog. They also have the ability, unusual for conifers, to produce upright sprouts along their roots. In this way they spread without benefit of seeds.

Under the thin shade of the tamaracks, black spruce seedlings eventually begin to appear. These, too, can produce new roots on their stems, as well as on lower branches that become buried. As the rooted branches in turn produce upright shoots, the mother tree soon has a flock of little ones growing around it. Spruces grow very slowly in the soddenness of a bog mat, and although a century-old spruce growing on dry land is usually about forty to sixty feet tall, one seventeen-foot tree growing in a bog was found to be 137 years old.

Black spruce has one of the most narrowly spirelike tops of all trees and gives a very characteristic pattern to a skyline. Cones are borne only at the very tops of the trees and behave rather like those of jack pine, shedding their seeds a few at a time over a long period, unless a fire comes along. Then they promptly open wide and release all the seeds that are left.

In the shelter of the tamaracks and along with the spruces there is often an invasion of northern white cedar. This tree is also known as arborvitae, a name that supposedly came from the fact that the Indians used a decoction of its leaves to prevent scurvy. It is a tree with a will to live, if this can be said of a tree, for its powers of vegetative increase are remarkable. Like black spruce, it spreads by the rooting of its lower branches. When it tips over, as it frequently does in a bog, it often goes at a slow and stately pace, continuing meanwhile to grow upward at the tip. Over a period of time this may give a decided curve to the trunk. When the tree finally goes all the way over, the branches that happen to be on the upper side begin to grow as independent individuals; and as these in turn produce offspring by rooting of their lower branches, the resulting thicket can become both large and dense.

White cedar grows not only on the extremely wet peat of bogs but also on uplands, even steep cliffs that are always well drained and sometimes very dry. This long puzzled ecologists, until experiments confirmed the suspicion that there are two genetically different strains of this tree in the Middle West, each inwardly adapted to a distinct kind of habitat but the two showing no outwardly visible differences.

Wherever it is found, the northern white cedar is usually associated with limy soil. Since bogs are typically acid, this seems an odd situation. However, even in a bog, water draining in from the upland has dissolved in it lime and other minerals that are present in glacial deposits of the region. But the mineral-rich water at the edge never reaches the mossy mat in the center, since it is in the nature of bogs that there is practically no circulation of the water in them. Except at the outer fringes, the water is essentially rainwater, absorbed and held where it falls on the sponge-like surface.

Since pure rainwater contains practically no dissolved matter at all, a bog mat is a very sterile substrate. Although there is a good deal of nitrogen present all told, it is incorporated in the tissues of plants, living and dead, and until it can be released by the process of decay it is totally unavailable for the growth of other plants. This had led to the assumption, plausible but still unproved, that the insects captured and digested by sundews and pitcher plants serve as an important source of nitrogen for these small meat-eaters.

It is quite clear that the growth of bog plants is constrained by lack of available minerals. Parts of the vast muskeg in the bed of Glacial Lake Agassiz have been experimentally treated with fertilizer. Wherever nitrogen or phosphorus was added, there was a marked increase in the growth of both black spruce and heath shrubs. It seems likely that most "typical bog plants" have no special predilection for living in bogs but grow there simply because they can make do, however meagerly, in a place that is too poor to support the more demanding plants that in a better situation would offer them strong competition.

Various attempts have been made to find out how fast peat accumulates in a bog. This is complicated by a number of things. The older peat becomes tightly compressed by the weight of material lying over it, and the oldest part of it disintegrates into fine-textured, formless muck. Also, a certain amount of mineral sediment is mixed in with the organic matter. For long range rates of accumulation, the best test is to determine the radiocarbon ages of samples taken at various levels below the surface. This gives only average values, however, and it tells nothing about recent events.

To find out about the current rate of build-up and to assess the contributions of the various peat-forming plants, one patient, long-suffering, and ingenious student picked apart and weighed the components of a mass of peat taken from the top of a bog in Itasca Park. These he separated into sedges (64 percent by weight), mosses (7 percent), inorganic and finely decomposed organic matter (25 percent), and "unidentified" (the rest). As a measure of the current rate of upward growth, he used tamarack seedlings of various ages. He found that when these are pulled up carefully and disentangled from the rest of the peat, the stems show a series of small, ringlike scars, each of which marks the place where the scales of the winter bud fell off as growth started one spring, the interval between rings marking a year's growth. Since seeds when they first sprout lie right on the surface, the first ring of scars marks the level of the peat the year the seed germinated. From calculations based on this tamarack chronometer, it appears that the top four inches of peat were added at the rate of about half an inch per year, or twenty-four years for the top foot.

What birds and beasts will live in a bog depends largely on the nature of the plant cover. Song sparrows and red-winged blackbirds, minks, muskrats, and a variety of shrews and mice, as well as frogs and toads, are all common.

In winter the cedar thickets of bogs and other damp places provide food and shelter for large numbers of deer. Such "deer yards" may be busy places during severe weather, for as the cutover forest lands have grown up to brush, the deer population has increased enormously. At the same time the amount of available winter range has been much reduced. Consequently what there is, is usually crowded. As the animals mill around among the cedars their slender hooves churn and compact the soil. This seems to influence the shrub layer more than it does the trees, and the understory takes on a less boggy and more woodsy appearance without particularly affecting the trees.

During the drought periods that are frequent in late summer in this part of the world the top of a bog mat becomes extremely dry. Fire sweeping into the bog at such times may not only kill the shrubs and trees growing on it but may also burn deeply into the mat itself. When the water eventually rises again, it fills the partly emptied depression, and the bog reverts to an earlier stage of development. If the water is fairly shallow, the new vegetation that comes in may be like that of an open marsh, with cattails, reeds, and alders instead of the heath shrubs that grew there before. If the fire burns only superficially, a bog may be converted to a flat, wet sedge meadow. Where black spruce is present, a moderate fire will make the old cones open and touch off a new generation of seedling trees.

In the natural course of events the ultimate fate of a bog is obliteration. This may come about extremely slowly as an outlet stream gradually develops and the bog drains; but filling with peat and other sediment is faster and much more common. As the water becomes shallower and some slight drainage movement begins, aeration improves and the water temperature rises. This causes a gradual shift from bog vegetation to the maple, ash, and elm of a swamp forest. Then only the presence of a deep layer of peat underground reveals the fact that a true bog ever existed in that place.

The conversion has also been known to go the other way. An instance of this was observed when a falling water table dropped below the saddle between two hollows and deprived a freshwater lake of its outlet stream. After this event the now undrained swamp gradually turned into a typical sphagnum bog.

Since the first settlers moved into the Middle West, enormous areas of former bog and muskeg as well as freshwater swamp have been artificially drained. In a wooded bog the most immediate effect of drainage is an abrupt increase in the growth of the trees. If written records are lacking, the date of a drainage operation can be read clearly in the widened growth rings of the trees. The increased air supply also causes a rapid oxidation of the peat, which is reduced in the process to a fine-textured black muck. This has much less bulk than the peat from which it is formed, and the land surface falls accordingly. The muckland is soon invaded by a new set of plants, among which alder and white cedar are conspicuous. In fact, it is suspected that many of the dense cedar thickets that are now found growing on damp, mucky peat originated as a result of widespread drainage and a general drop of the water table.

Drained peatlands have developed into some of the richest tracts of muck cropland. Any intensely black soil lying on the flat bottom of a low place probably marks the site of a former bog or swamp. Not all drained peatlands have proved to be profitable for cultivation, however. If the original peat was shallow, or if it lay over rocky or sandy soil, attempts to farm it have usually ended in failure. Another limitation is that cleared bogs are often subject to frost — for the same reason that the place was a bog to begin with. This is a greater hazard in the more northern regions, though; and in Ohio and southern Michigan some very prosperous truck farming is based on rich, black mucklands that once were bogs.

In the first decades of this century many grandiose drainage projects were undertaken in the upper Lakes States, with the hope of developing good farmland in place of the large areas of unusable muskeg. A few of these came through successfully; but much of the reclaimed land was never really worth cultivating, and many towns and counties that sponsored the projects had a hard time getting out from under the debts they had so hopefully incurred. The best that came of many of the projects was a marked improvement in the growth of existing bog forests. This benefit was short-lived, however, for the next generation of trees after cutting consists largely of alder and red maple, and these are not worth much economically.

Most of the more southern bogs have already disappeared, either by man's intervention or by natural processes, and the survivors exist in detached and misplaced climatic pockets. In the long enough run, even the northern muskegs can be expected to succumb to natural filling and ultimate draining. As for the upland spruce forests, what will happen to them seems to depend on the imponderable factor of climatic change.

9. The Edge of the Forest: Land of Oak

One of the most pleasing prospects that primeval America ever offered must have been the country that lay along the border where the woodland met the prairie. Here the dark forest edge lay in irregularly sweeping curves, now indented with rounded embayments, now sending dark capes and peninsulas out onto the grassland. Beyond the forest itself there were islands where the broadly rounded heads of the open-grown oaks were grouped in groves or clusters, or sometimes scattered singly like the trees of an orchard, or stood as outlying sentinels far out on the open prairie. This was the land of the groves, the "oak openings" that so charmed all who saw them in their original wild state.

In this borderland one had the best of two worlds. The shade and shelter of the spreading trees broke the fierce glare and heat of the summer sun and tempered the searing blast or the wintry bite of the prairie wind; yet between the trees was the arching sky above, and here and there a long view carried the eye out and away over the wide grassland. For the early settlers such places were ready-made homesites, and the prairie groves were taken up at an early date. Besides their beauty they had the very practical virtue of often standing on slight swells or ridges and from winter to early summer forming dry islands in an otherwise marshy expanse.

By far the most frequent tree of the groves and openings was the bur oak. There were other groves of white oak and even a few of hickory, and where the soil was dry and sandy, black oak was common. But all these were greatly in the minority, and an oak opening most often meant bur oak.

Compared to forest giants, the trees of the prairie groves were not very large. Trunks a foot thick would be typical. It was the form of the spreading crowns and the gnarled branches that gave the trees their special character.

One of the charms of the openings was the grassy carpet that covered the ground below. Although there was some variation according to the density of shade, the usual thing was a thick sod of prairie grasses, with here and there a thicket of shrubs. In the predominantly bluestem greensward grew and blossomed a long series of flowers, from birdsfoot violet and the bright yellow puccoon of springtime through the softer hues of wild geranium, roses, and wild bergamot to the late summer sunflowers and flowering spurge.

Except as a source of firewood and an occasional post or prop, the gnarled trees of the oak openings were of little use to a farmer. Consequently the groves were cut only where space was needed for house and farm buildings or to round out the area of a plowed field. In that part of the country practically all of the woodlots that remain today originated long ago as prairie groves, and around some of the older farmhouses a few widely spreading trees remain from early times.

The oaks of the openings were outriders of a forest of oak and hickory that lies across the center of the country in a broad belt from eastern Texas to southern Wisconsin. From the main body of the forest, long arms branch out in various directions. One rather wide arm reaches northeast across Indiana into the maple forest of southern Michigan, with spotty extensions as far east as central Ohio. Others run northwest up the rivers of Missouri, Iowa, and Nebraska. Yet another long, narrow strip runs northwest from Illinois over the nearby corner of Iowa and cuts diagonally across Minnesota, forming a buffer zone between forests of maple or pine and the grasslands of the west.

The heartland of this oak forest, where it is best developed and where it has lived for the longest time, is the hilly Ozark country of Arkansas and southern Missouri. In this homeland it has lived continuously since it evolved from the world-encircling temperate forest of Tertiary times. From there it has expanded and retreated as ice sheets ebbed and flowed and as climates have changed over the ages. There is reason to believe that it may also have lived on in the Driftless Area of Wisconsin during the Pleistocene glaciations, spreading out from there as the ice withdrew to the north.

There is a paradox about this forest, for on maps that show the distribution of the different vegetation types, the parts of Illinois and points eastward that are shown as bearing oak forest are also included in the prairie grassland area. Since the region of oak-woods-cum-grassland protrudes far eastward between pure forest on the north and pure forest on the south, it has become known as the Prairie Peninsula.

A large and detailed map of the Prairie Peninsula shows that it is really a mosaic of prairie and forest. Toward the west, in Illinois, the open prairie lies — or more properly lay — in large, horizon-wide expanses. Farther east the grassland tracts became smaller, until in central Ohio there were only scattered bits of an acre or so to a few square miles in extent. The savannah landscape of the oak openings represents the other face of the mosaic, where woodland tracts became smaller and finally thinned out to single trees deployed upon the prairie.

Even a large and well grown oak forest with a continuous leafy canopy overhead had an openness about it. The light is rarely so dim as it is in a maple woods, and there is usually a fair amount of undergrowth. Hazel, gooseberries, and brambles may form a dense shrub layer, or Virginia creeper and poison ivy may run along the ground and climb high into the trees.

Where the shrubs are sparse and more light reaches the forest floor, there is an abundance of small flowering plants. In spring the oak-leaf carpet is delicately flecked with rue anemone or plumed with false Solomon's seal. Later on come the more colorful large-flowered bellwort and wild geranium, while summer brings the taller growing lavender bergamot and spreading dogbane with its pink coral bells. As in any part of eastern North America, summer's end is heralded by the opening of asters and goldenrods.

The ground beneath the plants is covered year-round with a thick layer of leaf litter, for it ordinarily takes several years for a season's crop of fallen oak leaves to decay.

The oak forest provides a living for a wide variety of animals and is usually a rather lively place. Bluejays, cardinals, chickadees, and the various woodpeckers are birds with very unretiring habits, and the strong voices of woodthrush, veery, and oriole resound through the woods. A rustling in the brush is likely to come from a towhee scratching noisily among the dry leaves on the ground. Once there were also flocks of wild turkeys, but these are long gone from the Middle West. And the passenger pigeons that once darkened the sky for miles have been wholly extinct since 1914, when the last solitary specimen died of old age in the Cincinnati zoo.

The brushy undergrowth offers browse and shelter for deer, rabbits, and woodchucks as well as homes for raccoons and opossums. The large seed and nut crop feeds a host of mice, chipmunks, and squirrels; and foxes in turn make a good living from the large rodent crop.

Among the Midwestern oaks the cast of characters is rather large; but even an unfamiliar one can usually be recognized as some kind of oak. The leaves are typically dark and glossy, with a hard texture that is more woody than leathery. Trunks and branches are solid, strong, and

U.S. Forest Service

The oak forest is open and cheerful, with an abundance of undergrowth.

sturdy looking, even when they do not become picturesquely gnarled. Winter buds are clustered at the tip ends of the twigs, each one enclosed in a series of overlapping scales arranged in five rows. They are among the last to leaf out in spring; and traditionally, when the oak leaves are as big as a squirrel's ear, it is time to plant corn, for by then the weather is mild and settled and the chance of frost is practically gone.

Compared with other types, the oak forest is rather unstable and constantly changing. None of the oaks can reproduce successfully in its own shade, so they tend to be one-generation trees, shading out their own offspring and allowing others to take over. Judging by their relative shade tolerance, one might expect bur oak to be replaced successively by black and then white and then red oak. But although the red is the most shade-tolerant of the oaks, a number of other trees are much more so. Consequently there seems to be no remote possibility that oaks could ever develop the kind of self-perpetuating dominance that sugar maples are capable of.

In most directions, except for the complication of the Prairie Peninsula, the oak forest merges into forests of other types; but westward it meets the prairie. Its final decline begins just beyond the Mississippi and has three different manifestations: a reduction in the number of species as one after another reaches its western limit, a reduction in the size of the largest trees, and a restriction of the forest to the most favorable growing conditions.

From Ohio to eastern Iowa the oak forest is much the same. There the dominant trees are red and white oaks, with lesser numbers of such species as shagbark hickory, sugar maple, and bur oak. In the lower part of the forest canopy the more delicate leaves of hop hornbeam lighten the heavy texture of the oak foliage.

West from the Missouri River, one species after another reaches the edge of its range. In eastern Nebraska the maples and white oaks are gone; and only a little farther west, near Lincoln, red oak and shagbark hickory drop out, leaving the upland forest to the bur oaks and bitternut hickory. Beyond this there is little on the upland that could be called forest at all.

Although individual oaks growing in favorable situations can reach heights well over a hundred feet, they rarely grow more than eighty or ninety feet tall in places that are dry enough for the oaks to be dominant. This is what one finds in central Illinois or along the Missouri in eastern Nebraska. Around Lincoln forty feet is a good height for an oak woods. These smaller trees are rather widely spaced, making a more open woodland, and their trunks branch more and produce more widely spreading crowns than those of denser stands.

At the extreme edge of the forest, the last tree to give up is usually the bur oak. Although this persistent species can grow 170 feet tall in an Indiana bottomland, it peters out at the prairie's edge to a brushy shrub. It commonly forms a patchy, chaparral-like border along the entire edge of the woods, and there are clumps and mounds of it scattered far out on the open grassland. In the north the brush usually includes quaking aspen and balsam poplar as well as bur oak. Elsewhere prairie crab and wild plum form low, dense mounds, filled in springtime with fragrant bloom.

Most of the shrubs that can hold their own against the prairie, and sometimes even expand into it, produce wide-ranging underground stems. These work along a few inches below the surface of the soil and at intervals send forth clumps of roots and upright shoots or suckers. The suckers can live and grow indefinitely by drawing on the resources of the parent plant to which they are attached. This gives them a great advantage over seedlings of any kind, for they can rapidly push their leaves above the shade cast by the dense grasses.

Since the suckers of most prairie shrubs arise at rather short intervals, the brush expands by a slow but massive invasion along a solid front. Sumac, however, advances by leaps and bounds. Its underground stems run along a little below the surface for long distances, and new upright shoots appear as much as twenty or thirty feet beyond the parent plant. The main roots of a sumac clump grow down and outward for seven or eight feet, branching profusely as they go. Some of the branch roots then turn obliquely upward and as they near the surface they develop a network of fine absorbing rootlets. This abundance of superficial feeding roots makes it possible for such plants to live in competition with the efficient and voracious grass.

In more equable climates, tree seedlings eventually appear in the shelter of a mass of shrubs. In time some of them begin to overtop the brush, and sooner or later a woodland develops. But along the prairie-woodland border, the only kind of tree that has much success against the grasses and shrubs is the bur oak, and even it may never grow bigger than shrub size.

When a bur-oak acorn sprouts, it starts off almost explosively. Even before the first leaves unfurl it produces a tap root about nine inches long. This easily reaches a length of three feet in its first year. In clean-cultivated soil, experimental seedlings have produced root systems five feet deep and thirty inches wide by the end of their first growing season. At the ripe age of three years, the roots were eight and a half feet deep with a spread of over five feet. This performance must indeed offer a challenge to other plants, even those of an unbroken prairie sod.

It offered enough of a challenge to a group of ecologists at the University of Nebraska to induce them to undertake the long, hard labor of excavating the entire root systems of some well-grown bur oak trees. The trees they worked on were part of an open stand on the prairie not far from Lincoln. The one studied in most detail was sixty-five years old, thirty-seven and and half feet tall, and fourteen inches thick at the base. Its vital statistics were impressive enough to warrant giving some of them here.

In the first fifteen inches below the soil surface, the taproot produced fifteen major branches. These ranged from two to seven inches thick. The twelfth root down, eight inches below the surface, was four and a half inches thick at its base and more than an inch thick at a point thirteen feet out from the taproot. At that point it lay thirty inches below the surface of the ground. The taproot itself went straight down for more than fourteen feet.

Some of the largest branch roots were traced for forty or fifty feet out from the trunk, one of them to over sixty feet. This was well over twice the spread of the tree's crown. The main branches of the taproot in turn produced many secondary branches, a number of which grew vertically downward as "sinkers." A few threadlike branchlets were traced to a depth of fifteen feet. The entire top six feet of soil were thoroughly filled with the roots of this tree, and the ultimate rootlets formed a fine meshwork just below the surface. This impressive root system could draw water from a large volume of soil and challenge the grasses on their own terms.

An unbroken prairie sod is virtually impregnable to any seedlings less aggressive than those of a bur oak. The dense grass foliage casts a heavy shade, and the fine, dense roots that fill the soil like a sponge to a depth of several feet absorb most of the moisture as fast as it percolates from above. In summer there are frequent periods when a prairie soil contains no available water at all.

On steep or broken surfaces, however, the grass plants grow less densely. This may account for the fact that trees and shrubs often appear on the slopes of rapidly eroding gullies in the prairie. It may also help to explain the presence of woodlands on the faces of steep scarps all through the prairie and plains country.

During wetter phases of the ever-turning climatic cycles, there is water enough and to spare, and many new trees manage to become established in the edge of the prairie. But a good start is no guarantee of a long life, and in the inevitable times of severe and prolonged drought that follow, the mortality rate is high. This has been observed repeatedly and was documented in detail in the catastrophic drought of the 1930's. Surveys made after the rains finally returned showed that in western Kansas between a third and a half of the native trees of all kinds were completely dead and many more had been badly injured. This included even the doughty bur oak.

Losses were much higher among trees planted in windbreaks, hedgerows, and timber claims. The last were tracts planted as a result of the well-intentioned but not very productive Timber Culture Act of 1873. This gave a farmer full title to an extra quarter section (160 acres) of land if he planted and maintained 10 acres of trees on it. In the usual planting, trees were set out six to nine feet apart in two directions, and for the first eight or ten years the soil was cultivated to tear up the grass and weeds and loosen the surface so that rain and melting snow would penetrate. However, the kinds of trees that were planted depended more on what was available cheaply from nurseries than on what was likely to succeed, and every dry period took a heavy toll. The drought of the 1930's practically wiped out what little remained alive on the timber claims.

Trees planted as windbreaks and shelterbelts have done somewhat better. Those that survived the drought years best were red cedar, hackberry, bur oak, and black locust. Here, too, for any chance of success the soil must first be plowed. The trees must be big enough so that their roots will reach below those of the grass and weeds. Then they must be cultivated until they cast enough shade to hold down the competition. Mulching the ground while the trees are small is very helpful, and sometimes the plantings are even irrigated.

After the trees are well established and begin to produce seeds, seedlings sometimes appear; but these are confined to the shelter of the grove and practically never invade the prairie outside it.

Around the edges of the Prairie Peninsula, in the country of oak openings, the border between forest and grassland was very different from all this. Here the trees grew to full size right to the edge of the prairie and even out on it. As the region began to fill up with settlers, and roads and cultivated fields replaced the primeval vegetation, a drastic change came over the openings. Within a decade, it seemed, the prairie sod between the old oaks grew up to a woody thicket as though sown with dragon's teeth. Commonly the new brush consisted of black oaks, although there might be other kinds as well.

All these young shoots were not seedlings, as one might expect, but clumps of sprouts from stunted underground rootstocks. The pioneer farmer engaged in the back-breaking and often enough spirit-breaking toil of the first plowing of virgin prairie sod knew these large lumps of woody hardness all too well. He called them "grubs" and he cared little whether they were black oak, bur oak, redroot, or anything else. At the first plowing he swore at them and let his bucking plow jump over them. Digging them out one by one with a grub ax was left for slack seasons on the farm in later years.

These woody grubs, or burls, were the product of fire. In autumn and early spring, the seasons when plants are dormant and dry but not protected by a blanket of snow, fire was a universal possibility and a frequent actuality on the open prairie. Any tract was swept by fire every few years, if not annually, and the fires burned on into the grass carpet of the prairie groves. The only plants that can survive very long under such a regime are those that die back to the ground every winter, like the grasses, and those that are large enough to stand above the reach of the flames and have a thick, fire-resistant, and insulative layer of bark on their lower trunks, like the oaks. Although objective data seem to be lacking, all observers attribute the long and universal survival of bur oaks to the thickness of their rough bark. Black and white oaks can endure the heat of more moderate fires, but they are often killed back to the ground by fire that has little effect on bur oaks.

The way typical grubs arise was demonstrated accidentally some years ago in the arboretum of the University of Wisconsin. In 1947 a fire there burned through a stand of ninety-year-old Hill's oaks, killing them down to the ground. The root crowns sprouted again, but two more fires in the next five years finally burned up the last traces of the old trees as well as the more recent sprouts. Twelve years later Curtis reported that "these plants are now typical grubs again, with no surface indications that they are at least one hundred years old. Indeed, there is no good reason why they may not be a thousand years old, since the same process may have been repeated a number of times since the germination of the original acorn."

Over a period of many years the soil of a prairie grove became studded with grubs of widely differing ages. Between fires they all produced sprouts that were the same age above ground, giving the appearance of even-aged trees so typical of the oak openings. When the spread of settlement put a stop to wildfires entirely, all the hidden grubs sprouted at once, and the open groves promptly grew up to thick woodland.

Now that fires have all but ceased, the remnants of many

old groves are being preserved as such not by fire but by grazing. In all too many cases overgrazing has reduced the ground cover to a sparse stand of common weeds; but where the cover is thick and grassy such groves give a good impression of what the originals must have looked like. The plants in the undergrowth, however, are of very different kinds. Under grazing, the original bluestem grasses and other prairie plants disappear rapidly and are replaced by bluegrass and the many familiar weeds and field flowers that accompany it in pastures farther east, as well as in Europe, where they all came from. An oak opening complete with its original ground cover is probably as totally extinct now as the passenger pigeons that once roosted there.

There is other evidence that fire may have held the woods in check. A large proportion of the old groves and woodlands in the Prairie Peninsula stand in situations that afforded some kind of natural protection from fire. Often they are almost completely surrounded by water, lying in the loops of a meandering stream or on a ridge that rises through a once marshy prairie. Along any river wide enough to stop a grass fire the area of old woodland on the east or downwind side is two or three times the area on the west; for in clear, dry weather it is from the west that the high winds blow that drove the flames of a prairie fire.

Nobody seriously questions that the groves and the openings owed their nature and even their existence to the frequent occurrence of fire. The even ages of the trees, the dearth of underbrush, and the carpet of grass are typical results of repeated fires. The rapid resurgence of trees in the groves and the spread of trees onto the prairie as soon as fires were stopped also point to the influence of fire. The question then arises whether the prairies of the Prairie Peninsula would have existed at all were it not for the fires.

To this question there is no simple answer. In the upper Lakes States, the dry, sandy places that were burned over repeatedly and often usually bore pine or oak barrens or occasionally sedge meadows; but these did not turn into prairie. Far to the south and southeast, also, where the climate is warmer, recurrent fires have always influenced the vegetation; but here the effect is to perpetuate the pines and retard the invasion of hardwoods. On the coastal plain fire did not then and does not now produce grassland. The question remains: If not fire, then what?

An alternative possibility is climate. In pursuit of this idea, almost every conceivable aspect of climate has been studied. Temperature can be ruled out, since forests grow both far to the north and far to the south, as well as on the east of the Prairie Peninsula. Many other parts of the world with the same amount of precipitation bear a continuous forest. But there is more to the question of water supply than how much rain and snow fall in a year.

Back near the turn of the century the botanist Edgar Transeau pointed out that not all the water that falls on the earth is thereby available for the growth of plants, since some of it evaporates from the soil before plants can take it in, and most of what is absorbed evaporates rapidly from the broad surfaces of leaves. In a dry environment, a plant needs large amounts of water just to replace what it loses by evaporation, let alone grow. So Transeau undertook to unravel the relationships among rainfall, evaporation, and plants.

The maps he constructed show that wherever the year's precipitation equals or exceeds the total potential evaporation, the land bears some type of forest. Where evaporation is proportionately greater, and rainfall is only 60 to 80 percent of what could evaporate, the land bears tallgrass prairie. The region between, where potential evaporation just slightly exceeds actual rainfall, is the land where prairie and trees blend in the oak openings and where the oaks form a dry, open woodland.

Another aspect of climate that correlates fairly closely with the Prairie Peninsula is the seasonal distribution of moisture. In this region as well as in part of the open oak forest country, a good share of the year's precipitation falls

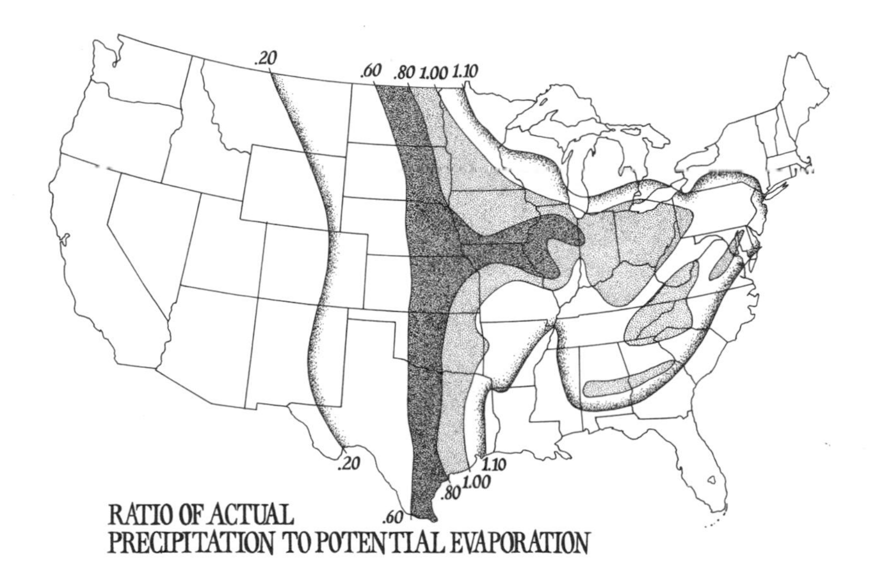

during the growing season, and autumn and winter are rather dry. The wet season usually ends by midsummer, and later summer tends to be droughty, even in normal years.

All this suggests that the grasses and their companions gain a significant edge over the trees from their ability to survive through the generally dry fall and winter and through an occasional protracted drought, along with their tolerance for a dry atmosphere and resistance to frequent burning.

Although the advance of the forest onto the prairie went on at a spectacular rate in the early years of settlement, there is abundant reason to believe that it had already begun before agricultural man appeared on the scene. One kind of evidence for this appears in the soil underlying the forests. East of the prairies patches of beech-maple forest have long been known that are growing on typical prairie soil. Along the prairie-forest border, the last mile or so of woods often stands on the kind of deep black soil that forms only under a cover of grass. A comparison of detailed soil maps with vegetational maps based on the pre-settlement land survey shows that this was true even under undisturbed, primeval conditions.

The University of Illinois is right in the heart of the Prairie Peninsula, and the forest and prairie thereabouts have been studied by biologists there for half a century and more. Everything that has been learned in that time confirms the general ideas proposed long ago by Professor Henry Allen Gleason. He holds that on the Illinois prairie the trees have been advancing for several centuries from the forest country in a generally northwestward direction, primarily along stream valleys, where the influence of fire is curbed.

In recent years ecologists at the university have turned their attention to a detailed study of the interaction between vegetation and soil. Several of their study areas in the central Illinois prairie country offer evidence that supports Professor Gleason's ideas.

One of these is at Allerton Park, along the Sangamon River southwest of Urbana. This is a typical sample of the forest that follows the larger valleys of the prairie region. A part of it is almost certainly in a virgin condition. This is a well grown forest of oak and hickory that rises to the moderate height of seventy to eighty feet. White oak is the commonest tree, but there are also some red and black oaks. The forest is rather open and includes such relatively light-requiring trees as black cherry, hackberry, and sassafras.

This is a mature and stable forest and it gives no indication that it is changing in any significant way. The soil beneath it is typical of oak forest anywhere, with a well developed profile, a rather light color, and only moderate fertility. It shows no sign that it ever was influenced by grassland. If it was, it must have been an exceedingly long time ago.

A few miles northeast of Urbana lie Brownfield Woods and Trelease Woods, two remnants of what was originally a large prairie grove lying in a bend of the Salt Fork River. Brownfield Woods had been a favorite picnic spot for people in the vicinity ever since the region was first settled, and some took a dim view of the change when the university took control of the area about forty years ago for purposes of scientific study, which meant excluding the public entirely.

Brownfield Woods consists predominantly of sugar maple, which as usual is reproducing itself in large numbers. There are also a few red oaks and a scattering of basswood, hackberry, bur oak, and the like. The trees reach up to a tall hundred feet or more.

One of the largest trees is a giant bur oak, with a trunk over five feet thick and fully equaling the hundred-foot height of the largest maples. It is estimated to be a good four centuries old. Perhaps the most interesting thing about this striking tree is that the lower part of its trunk bears the scars of large limbs such as develop only on trees growing in open array, as they do in the oak openings or on the open prairie.

The soil beneath this woodland tract shows the layered pattern that is characteristic of forest soils in general, but it is not very distinct, and its entire thickness is almost as dark-colored as prairie soil. It looks like a stage in the transformation of soil that had long been under the influence of grass but has recently become covered with trees. The large open-grown bur oak adds to the likelihood of this history.

The nearby Trelease Woods is growing on a deep, black, typically prairie soil that shows no evidence at all of the influence of trees.

These three pieces of woodland appear to represent three stages in the advance of the forest onto the prairie. Trelease Woods has only recently invaded what is still prairie soil. Brownfield Woods has existed long enough for the trees to have influenced the structure of the soil, but the stamp of the prairie remains strong and is confirmed by the ancient, open-grown bur oak. And at Allerton Park the oak forest has stood so long that the soil shows no evidence of anything else.

Another place that offers firm evidence that the forest was advancing long before the country was settled is northwestern Minnesota. A traverse westward from Itasca Park takes one, in the span of less than forty miles, from the tall forests of Itasca, partly pine and partly hardwoods, through typical hardwood forest to oak scrub and brush prairie and finally out onto the horizon-wide flats where nearly all the original open prairie is now converted to farmland.

The Itasca region shows all the signs of a forest of long standing that has been repeatedly burned: groves of pine, mostly even-aged but with a scattering of giants that survive from an earlier fire-begotten generation; a few large fire-resistant bur oaks standing out much like the giant pines; white birch of all ages; and often a grassy ground cover. The soil has the structure typical of pine forests of long standing. But in spite of all the evidence of past fires, the land remains forested.

As one heads west from Itasca, the last patch of dense hardwood forest before the brushy prairie border stands on a high moraine near the little town of Waubun. The dominant trees here are sugar maple and basswood, along with the smaller hop hornbeam. These are the same as the chief hardwoods in Itasca Park, but here near the forest edge they average five or ten feet lower in height. There are a few large bur oaks in these woods, but they are no bigger than the maples and basswoods, and their straight, unbranched trunks show that they grew up in the woods along with the other trees and are not leftovers from an earlier time.

Unlike the forest at Itasca, this one at Waubun is growing on a typical grassland soil, deep and uniformly dark except for a slight leaching of the very top part. Yet this forest must have been growing here since long before any human interference brought a halt to prairie fires. A few stumps of recently cut maple trees whose growth rings were still countable when they were examined a few years ago all dated from about 1810. But the region was not opened for settlement until 1906, and in 1810 that part of the country was still a remote, virtually unknown wilderness.

The structure of the forest itself suggests that it is probably in its first generation. There are no giant trees remaining from an earlier time, none of the paired mounds and hollows in the soil that are left when a large tree is blown over and torn up by the roots. There are no indications that the forest has been burned, although according to long-time residents fires were still common on the prairie a little to the west in the early years of this century. There are no old, open-grown trees, bur oak or other, and no pines at all. The western border of the woods is quite sharply defined. The whole situation looks like what one might find if, at some time upward of a century and a half ago, the maple-basswood forest took a giant step westward and migrated en masse onto the edge of the prairie.

One might, for the sake of symmetry, expect to find other places where prairie in turn is growing on typical forest soil; but this does not seem to happen. The only places where grassland has pushed into existing forest in recent times are a few where it can be shown for certain that the trees were destroyed by fire; and when the fires are over, the trees soon return to their old territory.

Evidence of a different kind, but pointing in the same direction, appears in the pollen records for this part of the country. Northwestern Minnesota has an abundance of bogs and ponds whose pollen profiles are particularly clear-cut and can be studied in fine detail. Many of the bogs are still undisturbed, and their records continue right up to the present.

These bogs show clearly that the forest has advanced onto the prairie over the past several centuries. They also show that the trees were merely returning to areas they had occupied before the warm, dry period that reached its peak about four thousand years ago. In Minnesota the total shift was a matter of some forty to sixty miles.

In the Prairie Peninsula the vegetational shift was much more dramatic. This entire large region seems to have a climate that is close to the borderline between one that favors grass and one that will support forests. The postglacial warm period was enough to tip the balance toward grass, and the prairie expanded far eastward. When the climate became slightly cooler and moister again, the trees began to return; but the change in climate apparently was not great enough to offset entirely the grass-perpetuating effect of the continuing prairie fires. When at last agricultural man and his fire-stopping works eliminated this last obstacle, the trees erupted suddenly all over a large area.

In Nebraska and Iowa, as in Minnesota, the postglacial shifts seem to have been rather small. This region lacks the abundance of bogs and ponds that one finds in the north, so the extent of the shifts has not been read in such fine detail. At present, the forest tapers down to brush along the prairie's edge in a way that suggests a fairly stable boundary. With the ending of prairie fires in this region the trees have spread rapidly along river valleys, but only on the damp bottomlands, and the broad prairie-covered uplands have never become wooded.

It might be that with the fires stopped, the Prairie Peninsula would have grown up completely to forest. But the suggestion is more or less academic, since nearly all the land that was not already in trees, and a good deal that was, is now entirely occupied by farms, cities, suburbs, and the highways that connect them all.

10. The Prairie

So DIFFERENT from anything in their European homeland was the wide sweep of unbroken grassland that once occupied the center of North America that the first explorers to look upon it had no name for such a thing. For lack of a better, they used the word "prairie," which in their native France meant simply a large meadow. Since the English brought no more appropriate word from their own domesticated land, they took over the one used by the French, and "prairie" it has remained to this day.

The vast, unpunctuated openness of the prairie seems always to stir a strong response in those who see it for the first time. To some it is repellent, a place of monotony and emptiness. Such was the English traveler in Illinois who reported: "I . . . crossed the river, and came for the first time into the large Prairies, which, from their size, almost entirely lose their beauty, and present nothing but an immense sea of grass." Another has spelled out his objections in more detail. "To the traveller, who for several days traverses these prairies and barrens, their appearance is quite uninviting, and even disagreeable. He may travel from morning until night, and make good speed, but on looking around him, he fancies himself at the very spot whence he started. No pleasant variety of hill and dale, no rapidly running brook delights the eye, and no sound of woodland music strikes the ear; but, in their stead, a dull uniformity of prospect spread out immense."

Almost universally, travelers over the unbroken, unfenced prairie likened the journey to voyaging on the ocean. It was not just the waves that ran before the wind through the tall grasses, or the largeness of the sky all around; but the tracklessness, the absence of landmarks, the resemblance of the edge of the forest to a shoreline and of a settled place to a harbor, or a prairie town to a seaport. As the same Englishman observed, "I do not know anything that struck me more forcibly than the sensation of solitude I experienced in crossing this, and some of the other large Prairies. I was perfectly alone, and could see nothing in any direction but sky and grass. Leaving the wood appeared like embarking alone upon the ocean, and, upon again approaching the wood, I felt as if returning to land. Sometimes again, when I perceived a small stunted solitary tree that had been planted by some fortuitous circumstance, I could hardly help supposing it to be the mast of a vessel. No doubt the great stillness added very much to this strange illusion. Not a living thing could I see or hear, except the occasional rising of some prairie fowls, or perhaps a large hawk or eagle wheeling about over my head."

Although some find themselves unsettled by the rolling vastness, others feel expanded and exhilarated by the long view, the largeness of the sky. Their eyes perceive detail and they respond with joy to the boldly bright flowers and the birds and the surging richness of the living grassland. Here is Hamlin Garland, speaking from his anything-but-idyllic childhood on the Iowa frontier. "In herding the cattle we came to know all the open country round about and found it very beautiful. On the uplands a short, light-green, hair-like grass grew, intermixed with various resinous weeds, while in the lowland feeding grounds luxuriant patches of blue-joint, wild oats, and other tall forage plants waved in the wind. Along the streams and in the 'sloos' cat-tails and tiger-lilies nodded above thick mats of wide-bladed marsh grass. — Almost without realizing it, I came to know the character of every weed, every flower, every living thing big enough to be seen from the back of a horse.

"Nothing could be more generous, more joyous, than these natural meadows in summer. The flash and ripple and glimmer of the tall sunflowers, the myriad voices of gleeful bobolinks, the chirp and gurgle of red-winged blackbirds swaying on the willows, the meadow-larks piping from grassy bogs, the peep of the prairie chick and the wailing call of plover on the flowery green slopes of the upland made it all an ecstatic world to me. It was a wide world with a big, big sky which gave alluring hints of the still more glorious unknown wilderness beyond . . . The sun flamed across the splendid serial waves of the grasses and the perfumes of a hundred spicy plants rose in the shimmering air." And again, "The west wind comes to me laden with ecstatic voices . . . Vast purple-and-white clouds move like stately ships before the breeze, dark with rain which they drop momentarily in trailing garments upon the earth, and so pass in majesty amidst a roll of thunder."

Repelled or enchanted, lover or loather, no one could witness unmoved the spectacle of a prairie fire. And in the primeval grassland world, fire was one of the constants. In early spring, after the snow was gone and before the lush new growth had pushed up through last year's dead remains, or in autumn after a hard frost had killed the season's crop of herbage and before a new blanket of protective snow had fallen, the luxuriant mass of dry foliage formed a continuous expanse of fuel. Where the ground was damp, or on the occasional day when the air lay still, fires were light and often moved idly, desultorily over the land. Then any small stream or scarp that broke the grass cover would stop the slowly creeping flames, and even a few plow furrows scratched around a homestead would protect it.

But fire driven by a strong wind over land that had been thoroughly dried by days of sun and high wind was another thing altogether. Then it swept along in a continuous wall that could be a mile or more long. "The line of flame rushed through the long grass with tremendous violence, and a noise like thunder; while over the fire there hovered a dense cloud of smoke. The wind, which even previously had been high, was increased by the blaze which it fanned; and with such vehemence did it drive along the flames, that large masses of them appeared actually to leap forward and dart into the grass, several yards in advance of the line. It passed me like a whirlwind, and with a fury I shall never forget."

A prairie fire by day was spectacular enough, but a disembodied mass of flame seen against the blackness of night was positively theatrical, especially when several separate fires could be seen at once moving rapidly over the field of view.

Sometimes a traveler on the open prairie had warning of the distant approach of fire. There might be a tinge of smoke smell in the air, or flakes of ash borne far in advance of the fire itself. Or a distant glow might appear in the sky or a column of smoke on the far horizon. Then his first thought was for immediate refuge, not only from the flames but from the possibility of being blistered at a distance by the searing heat. If a large river or pond was accessible, he might be able to put it between himself and the fire, or better still, take a stand in the middle of it where he could submerge himself and his possessions during the worst of the fire. Otherwise the best hope of safety was to set a backfire that would clear the combustible matter from a refuge area where traveler, gear, and animals could stay while the fire swept around and past them.

In the wake of a severe fire the entire landscape was left a dreary, ashen gray, blackened here and there with the charred remains of shrubs and grass tussocks. This was a familiar sight to those who knew the prairie in its early days. But when spring brought fresh growth to the unharmed roots and rhizomes of all the prairie plants, they burst forth with renewed vigor from the earth that had been freed from its thick overlying mulch and enriched with ashes.

In such prairie tracts as remain today, the round of seasons continues as from time immemorial. As soon as the first drying spots emerge from the universal wetness of the spring thaw, the long progression of growth and bloom and fruiting begins. Even before the greening of the first grass tips, on the drier slopes and ridges the fat, silky-gray buds of pasque flower push up and soon unfold into large purple blooms. As the soil on the lower places begins to drain and warm, the new foliage of grasses fills in a green background for a procession of flowers — white of anemones and adderstongue, yellow of buttercups and puccoon, purple and yellow of violets, birdsfoot, and others, and the purplish red of the picturesquely named prairie smoke. White clusters of strawberry blossoms give promise of fragrant fruit to come, and above the lengthening grass blades rise large creamy masses of false indigo blossoms and drifts of fuzzy white redroot.

In the chill of damper places things start more slowly, and early spring finds little activity there. But by June spiderwort and bright golden Alexanders appear, water hemlock stands tall, and blue flags unfurl in profusion along the sloughs.

The full tide of summer brings a new set of prairie flowers, many of them large and strongly colored. Coneflowers, blackeyed Susans, rosinweeds, and compass plant, all are robust members of the daisy family. Legumes number a host of species — vetches, wild clovers, milk-vetches, ground plum, and the lupine-like prairie turnip. Various gay feathers or blazing stars sort themselves out according to the degree of moisture that each one prefers.

As summer moves on toward autumn and the earlier bloomers are coming to fruition, asters and goldenrods in variety take over the main role of flowering. By this time the grasses are ripening their seedheads upon stalks that have overtopped all other members of the grassland. Both the seedheads themselves and the stems that bear them take on shades of red, purple, and russet brown. In the clear air of summer's end, backlighted by the low evening sun, the grasses have a luster and shimmer and a brilliance of color that the outlander rarely associates with the idea of "grass."

Probably the best times of year on the prairie are late spring and early autumn, for then the weather is likely to show its most temperate face. But throughout the year the prairie climate is given to erratic and unpredictable extremes.

One of the climatic immoderates is moisture. In the prairie country most of the year's precipitation falls in spring and early summer, fully half of it, in fact, in the three months from May to July. Late summer is often persistently hot and dry, with the sun grinding down on a parched land, while the air over the prairie shimmers with heat and buzzes with the cries of insects. Even this seasonal rainfall is not reliable, for while in some years it rains in moderation and at reasonable intervals all through the growing season, well over half of the summers have substantially less than the average or "normal" rainfall.

When it does rain, it is seldom in moderation. Protracted periods of light and gentle rain are rare, and the typical thing is a downpour or nothing. At least half the summer's rain falls at the rate of an inch in twenty-four hours or less, and often a single storm drops two or three inches.

Most of these are thunderstorms. When columns of rising air build up over the fiercely heated land and give rise during the afternoon to towering thunderheads, the dénouements that follow can be very spectacular. Heralded by squalling gusts from a blackening sky, the rain may come in a cloudburst, falling in such torrents that water stands an inch deep on the level. Bolts of pinkish yellow lightning dart through the lowering clouds, sometimes streaking so close together that they form an eerie network over half the sky and light up the landscape for a split second as brightly as noonday, while the thunder cracks and rolls, until at last the storm moves off into the distance with a diminuendo rumble.

Temperature in the grassland also tends to extremes.

Daily weather maps show that practically all winter the coldest part of the country is not far from the Red River Valley of Minnesota and North Dakota, while in summer the same region is one of the warmest, giving the Red River Valley the widest annual temperature range in the country.

It is in the prairie region of Kansas, Missouri, and parts of Iowa and Illinois that rapid changes in the form of heat waves and cold waves occur most often. When a severe cold wave bears down, a midwinter day that began with the air feeling damp and soft and everything streaming with meltwater may change in a few brief hours to a screaming blizzard with the mercury well below zero. Or the long grind of summer's heat may end abruptly for the year when the first mass of cold air moves down from the Arctic, making the fretful sleeper relax in the blessed coolness and finally stir himself in a sleepy hunt for the blankets that have been so long unused.

At any season of the year the grassland climate is windy. Over the level midregion of this continent, from the Arctic Ocean to the Gulf of Mexico and from the foothills of the Rockies to the rough land near the Ohio River, there is nothing to interrupt the continental circulation of the air and little to deflect the winds from the surface of the land. In early summer the prevailingly southerly winds are warm and moist with water evaporated from the Gulf; but from other quarters the prairie wind is predominantly dry. From the west the air has been warmed as it descended from the mountains, where most of the moisture was wrung out of it. From the north it is warmed and dried as it travels overland by the progressively stronger rays of the sun. And although winds may blow locally from an easterly quarter, weather-sized air masses rarely if ever move from east to west.

The general dryness and windiness along with the predominantly clear skies give the prairie air a tremendous power of evaporation. Even an abundant rainfall evaporates very rapidly from the soil and from all the surfaces of vegetation, and it takes a correspondingly larger amount of rain just to keep the soil moistened. Add to this the fact that much of the rain falls faster than the ground can absorb it, so that a considerable amount of it runs off directly into the streams and rivers, and it becomes clear that "annual precipitation" does not tell all there is to say about the water available for plants to grow on.

The country of the tallgrass prairie extends west to approximately the hundredth meridian. Beyond that, as the land rises higher, the climate becomes progressively drier in the rain shadow of the Rockies, and the tall-growing grasses of the true prairie give way to the shorter herbage of the Great Plains.

Over all its vast extent, for hundreds of miles, the prairie landscape is dominated by the same small handful of grasses. From the one-time prairies of Ohio to the wide expanses in Nebraska, and from Canada across Minnesota and as far south as Texas, the chief characters are four: big bluestem, little bluestem, Indian grass, and switchgrass.

If one plant could be called *the* typical prairie plant, it would be the big bluestem grass, *Andropogon gerardi*. This handsome plant is by no means restricted to the prairies, for it grows over a large part of eastern North America, from Quebec to Florida and from New Mexico to Saskatchewan. But while in forested country it grows sparingly in the few open places, on the prairie it is abundant everywhere and commonly accounts for as much as three-quarters of the green cover of the land, often appreciably more.

Other prairie grasses are also widely distributed over the continent. Little bluestem (*Andropogon scoparius*), Indian grass (*Sorghastrum nutans*), switchgrass (*Panicum virgatum*) and slough grass or cordgrass (*Spartina pectinata*) as well as big bluestem all grow along roadsides and railroad tracks and in assorted fields and meadows in such unprairielike places as Connecticut.

Big bluestem takes its name from its size and from the

waxy bloom that covers its stems and leaves. By midsummer the leaves may be three feet long, arching gracefully over so that, although the plants may actually grow in clumps with much bare soil among them, the foliage makes a dense, continuous cover and casts a heavy shade on the ground beneath it. As grasses go, it starts growing rather late in the spring — mid-April in eastern Nebraska, for instance — and being in general a warm-season plant, it keeps growing right through the summer. Once started, it develops rapidly. By early July the leaves have reached their full size and flower stalks begin to show above them. After that the plant is easy to recognize by the fingerlike forking of its flower- and seedheads. When the flowers open and begin shedding their pollen, well along in August, the flowerheads often stand six or eight feet tall, half again that high in exceptionaly favorable places.

In dryish soil big bluestem is considerably smaller and less rampant, making it more subject to competition from other plants, and it does not tolerate waterlogging at all. But on

The tallgrass prairie is so completely cultivated that big bluestem is now most commonly seen as a "fenceline relict." From its typically forked flower clusters this grass is sometimes called "turkeyfoot."

fertile soil that is both moist and well drained, it is undoubted master of the prairie. A number of things account for this. Among them are its fast growth, its large size, and the density of its foliage. Only a few weeks after the seeds germinate, the plants produce the short side-shoots known as tillers, as well as long creeping stems or rhizomes; so even in infancy the plants form substantial little clumps. By the end of their first growing season, seedlings may be a foot or more tall and have roots a yard long.

Although a prairie full of big bluestem is exposed to the full light of the sun, the plant is remarkably tolerant of shade in all phases of its life. It can grow vigorously in as little as a tenth of full daylight, and small seedlings thrive even in the deep shade of their parents. Consequently, generation can follow generation in the same place.

In a well grown bluestem sod, the top few inches of soil are filled almost solidly with roots and underground stems. Roots are abundant to a depth of several feet, and many of them commonly reach six or seven feet. With such exuberance and such productivity, and considering also that the plant is highly nutritious and that both wild and domestic cattle prefer it to all others, big bluestem is indeed a paragon of prairie plants.

Its close relative, little bluestem, averages perhaps half the size of its queenly sister, but it can thrive in considerably drier soil. It grows in dense clumps that are usually six to eight inches across, leaving space between them for other kinds of plants. By late summer, when the ripening seedheads open into a profusion of tannish fuzz, the stems and leaf bases take on a characteristic bright reddish brown color that lasts all winter and adds a note of warmth wherever little bluestem appears.

Indian grass is large and robust like big bluestem, but it has a rather yellower green color and its leaves tend to stand more upright. It is less given to branching from the base and often appears as sparse clumps or even single stems. It thrives especially well on disturbed places, and an abundance of it is often a clue to frequent fire or flooding.

Switchgrass and especially slough grass are most common in damp places. Switchgrass is likely to be somewhat bigger than big bluestem, with rather broader and yellower leaves. It is easily spotted from midsummer on by the wide spreading arrangement of the rather thin, wiry branches of its flower- and seedheads. A good stand of slough grass grows to enormous size, and all parts of the plant are rather coarse and heavy.

Kentucky bluegrass was not a part of the primeval prairie. In fact, it is not even native to Kentucky but was introduced into this country by the earliest settlers. It thrives here in many situations, however, notably in its namesake state, and it has become rather widespread on the prairie. It is a cool-season grass, and the first green tinge of grassy places anywhere in spring is likely to be contributed by the early-starting bluegrass. Its strong point is its resistance to close cropping, whether by grazing animals or by mower blades; its vulnerabilities are a low tolerance for shade and a high susceptibility to fire. So, although it may become important in places that are rather heavily grazed or closely mowed, it readily dies out where grass fires are frequent or where tall prairie plants shade it.

All through the native prairie land there is a scattering of smallish woody shrubs. Such plants as redroot, sumac, and prairie rose are well able to hold their own among the dominant grasses. Often they expand by means of creeping roots and rhizomes into large compact clumps that may form quite conspicuous islands in the grassy expanse. But except near the forest edge or in the valleys of streams, the shrubs live in balance with the other prairie plants and in the long run tend to lose about as much as they gain.

Of all the plants of the prairie, it is the grasses in their millions that give the special character to the broad landscape. Right from its beginning a grass plant is well adapted to getting along in a rigorous world. As soon as the seed sprouts, the first root pushes downward so fast that by the time the shoot even begins to show green, the root has already reached a depth that is not likely to dry out immediately after a good rain. Even a minute seedling of timothy or redtop has a root two or three inches long when its shoot is just starting to swell, and a corn root penetrates five or six inches into the soil in as many days after it sprouts. By this time other roots begin to push out from the little nubbin of stem, and by the time leaves appear above the soil, a little grass plant is already endowed with a vigorous tuft of slim, fibrous roots.

Should even this sufficiency of roots fail to keep up with loss of water from the tops, and the plant approaches a state of wilting, the narrow leaves become folded or rolled, greatly reducing the surface exposed to drying wind or sun. If drought becomes severe, the entire plant may go dormant and simply suspend growth and most of its other vital activities for the duration. Even if all the leaves die of desiccation, the growing points and most of the mass of roots can survive in an inactive state for long periods of time, even several years, without actually dying.

The tendency of grass stems to remain close to the ground or even buried in it serves to protect the buds with their growing points from a number of hazards. Encased in the sheaths of older leaves and shaded by their overarching blades, they are shielded from extremes of heat and dryness of the prairie air. The low-lying buds also escape the worst effects of fire and the mouths of grazing animals. Add to this the fact that every grass leaf continues to grow at its base for a long time, even when its outer end is destroyed, and one can see why these plants can survive burning or grazing far better than almost any others.

Although the vegetation of the prairie is remarkably uniform over distances of hundreds of miles and over many types and textures of soil, it does vary somewhat according to the wetness or dryness of the land. All accounts of early days speak emphatically of the endless swampiness of the country — much of it all but impassable and a great trial to travelers and settlers. The oldest roads and houses were all located on morainal ridges or hummocks or on abandoned beach ridges of early postglacial lakes, for these were the only things that stood above the wetness on the great till plains and extinct lake beds of Illinois or Iowa. In much of the country, however, the standing water and the waterlogged soil lasted only through winter and spring, and by the time the weather was warm enough for the bluestems and Indian grass to start growing, most of the surface water had dried up or drained away.

On the best land big bluestem is overwhelmingly in control; but where soil moisture is less than ideal its vigor is strongly curbed and other grasses become increasingly abundant. In the wetter places, switchgrass and nodding wild rye are common, and in sloughs and swales, along stream edges, and on the "first bottoms" of larger streams, the tremendous and rank-growing slough grass takes over.

In drier situations big bluestem is also less vigorous, although even on quite dry prairies it still appears in practically every square yard. In such places little bluestem rises to prominence. It is generally more drought-tolerant than its big sister, perhaps because its finer, and if possible denser, roots are more efficient absorbers of water and its smaller leaves offer less extensive surfaces for evaporation. Other middle-sized grasses also come in on the drier places, but none of them approaches the abundance of little bluestem.

Like the dominant grasses, plants of the wetter prairie extend far into the forest country on the south and east; they are sometimes described as the Alleghenian meadow element. Those of the dry prairie, on the other hand, are rare or unknown farther east but range southwest into the still drier country beyond the prairie. It appears that as vegetation has migrated into the tallgrass prairie region since the end of the Ice Age, the special living conditions there have acted as a sort of filter, letting in only species from the east that can survive its special rigors and species from the west that can compete with them.

One of the striking characteristics of prairie plants is their ability to live so tightly intermingled, and in a soil where moisture is often deficient. The dense mass of grass roots and rhizomes that permeates the top few inches of the ground is further threaded through with the stems and roots of many other plants — goldenrods, sunflowers, mints of many kinds. There also bulbs of lilies and wild onions, tubers of the sunflower known as Jerusalem artichoke, and the thick, fleshy roots of a host of others.

Many studies have been made to assess the relation of this underground vegetative mass to the soil that encloses it. A number of these involve removing the soil matrix without disturbing the roots. All such procedures begin with digging a large, smooth-walled trench. Then the soil is removed from a sample block by a combination of streams or jets of water and laborious hand picking. Any method of root cleaning requires long and painstaking work; but more than one investigator has remarked in his report that the emerging pattern becomes interesting enough to make the labor tolerable. All told, hundreds of plants must have been studied in this way.

It is the grasses that fill the soil most densely. Under either bunch grass or sod, each square foot exposed contains hundreds of fine roots. Most of these lie fairly close to the surface, but a dense network extends to quite astonishing depths. On good sites the root mass of big bluestem or switchgrass commonly reaches a depth of seven or eight feet. In drier places little bluestem averages a more moderate four or five feet.

Roots of many other plants go even deeper than those of the grasses. Often a taproot grows straight down to below the thickest part of the grass roots before it branches at all. Leadplant, which easily reaches a depth of sixteen feet, makes its first branching at two or three feet. The modest-looking prairie rose may have a vertical taproot more than twenty feet long, the compass plant ten to fourteen feet, and even the rather small many-flowered aster more than seven feet.

On the other hand, a few prairie plants manage to survive with roots that seldom grow deeper than two feet. These include such small things as strawberry, catsfoot, and spiderwort.

All this means that there is much less direct competition among roots than one might suppose. The different plants take their water from different levels, and the vegetation as a whole draws upon a great thickness of soil. After a long period without rain, the soil may be completely dry; but when it does rain, the spongy, porous surface has a great absorptive power.

Such a mass of roots is bound to have a considerable influence on the soil it occupies. Each slender rootlet as it grows exerts a slight pressure on the material surrounding it, molding the soil particles into little crumb-like masses; and when it dies and eventually decays, it leaves a hairlike channel lined with a fine strand of its minute organic remains.

Dead and alive, the amount of organic matter in a prairie soil is very large. Under a good growth of big bluestem, the top foot may contain five tons per acre of organic matter, most of it in the top four inches. Although not all of this is current production, much of it consisting of roots that live for a long time, it still represents a substantial yield compared with the two tons per acre of the annual hay crop on the same land. This high organic content gives the soil not only its dark color, but more important its ability to retain water and, along with the channels formed by the many rootlets, its crumbly, porous texture.

The great fertility of prairie soil stems both from the influence of the prairie grasses and from the nature of its parent material. Its ultimate source is fine-textured glacial drift, whether till, outwash, or loess. Under the moderate rainfall of the region, there has not been time for much of the soluble minerals that serve a plant as food to be washed or leached away. The parent bedrock from which the drift was formed was well endowed with lime, and the resulting soil has not become acidified.

Grass plants, in contrast to such trees as oaks and pines,

absorb large quantities of lime with the minerals they take from the soil. This is transported upward into all parts of the plant, and as these eventually die and decay, their remains filter down into the root-filled soil. In the prairie climate the plants that subsequently cycle and recycle all this mineral wealth grow rather faster than the soil microbes can decompose their remains, and the darkly enriching organic matter continues to increase. The strongly blackened zone of prairie soil is commonly two feet thick, and in damp places it may be several times that.

A vegetation as productive as that of the prairie can support a large population of animals. Underground and in the thicket of stems and leaf bases just aboveground live a host of mice, gophers, ground squirrels — some with their famous thirteen stripes — and assorted other spermophiles. There also moles and badgers dig for a living. Prairie dogs, however, in spite of their name, cannot live in the luxuriant grass except where heavy grazing has reduced it to a close sward, as they do not seem to be able to cope with the problems of moving around and seeing before they are seen among taller herbage.

Even in "undisturbed" prairie, digging animals do a good deal of turning over of the soil. The large numbers of ground squirrels make extensive networks of burrows, and in the aggregate and over a long time, these little beasts move an astonishing amount of earth. Some plants are killed when they are buried under a mound of soil at the mouth of an animal burrow; but the soft soil increases the penetration of rainwater, and big bluestem only grows the more luxuriantly for the disturbance.

All these smaller creatures provide a living for the hunters — fox, coyote, and wolf. The predators have been so vigorously persecuted over the decades that the wolf no longer sings to the prairie moon; but the fox and coyote still survive.

Of the birds, perhaps the most uniquely characteristic of true prairie is the prairie chicken, now all too rare, which inflates his orange neck sacs and stamps out his dance on the stamping ground.

Although grassland birds are less numerous than those of the forest, they are much more in evidence. One hawk soaring in a clear prairie sky is visible to watchers for miles around, while the bobolink, meadowlark, and horned lark make themselves conspicuous by singing as they fly.

Prairie birds consume vast quantities of the abundant seeds produced by grasses and other plants. Another abundant source of bird food is the incredibly large number of prairie insects. Grasshoppers, leafhoppers, crickets, spittlebugs, beetles, spiders, ants, and other crawlers, runners, leapers, and fliers to a total of 1175 kinds have been listed for Iowa prairies. A good guess is that by late August each acre of prairie is inhabited by some ten million individual insects.

Living their largely separate lives were the grazers, chiefly the buffalo. These wild cattle once ranged by the millions all the way from the grassy meadows of the Ohio River and the Scioto far onto the high plains, beyond the Middle West. The explorer LaSalle is said to have seen a buffalo stuck in a marsh near South Bend, Indiana, early in the winter of 1679.

There has been much discussion over the years, most of it conjectural, about how much the tremendous numbers of grazing animals may have influenced the prairie. One extreme point of view holds that they may actually have been the cause of the treeless grasslands, rather like sheep within historical times in Great Britain. According to present knowledge, the buffalo wandered about in a very random fashion, without following any regular migration routes. Consequently, although from prehistoric ages until just over a century ago there were several million of these large beasts on the grasslands at any one time, their influence at any one place was only transient. Immediately after a large herd had passed through, the prairie would be closely grazed and severely trampled, with the soil churned up in some places; but in another season the effect would have passed, and it might be years before another herd came exactly that way again.

The more significant relation between grazers and grasses was probably a steady selective action of each on the other while both were evolving from their less specialized ancestors. The evolutionary history of the horse is known in by far the greatest detail; but similar changes affected the buffalo, and over a span of twenty million years they became more and more closely adapted to the grasses on which they fed.

Evolution of the grasses in turn was probably guided by the grazers. The short, creeping stems, the hidden buds, the leaves that grow on at the base when their tops are removed, the rough, hard texture that resists damage by pounding and trampling, all these traits would be the result of long-acting selective pressure. Whatever the cause, grasses and grazers now live in a state of close mutual adaptation, a thing that man has learned to exploit to his own great benefit; and the prairie that supported the wild herds continues to support man's domestic cattle.

In the Prairie Peninsula nothing at all is left of the prime type of tallgrass prairie. Farther west much land remains that has never been plowed but is used as pasture or mowed for hay. Many prairies are known to have been mowed every year for a century or more. This seems to have little effect on them, and many are still in a state very much like their primeval condition. Light grazing also leaves the prairie practically unchanged.

In fact, occasional mowing or light grazing usually benefits the vegetation by reducing the mat of dead litter that otherwise accumulates into a heavy mulch. If good prairie is neither mowed, grazed, nor burned, it may in a few years become covered with a six- or eight-inch thick mat of dead leaves and stems. This usually shades out any lesser understory plants and leads to almost pure stands of big bluestem, or sometimes switchgrass. It also acts as a very effective insulator, and the temperature beneath it may be twenty to thirty degrees lower than it is under unmulched grass. This may be beneficial in the heat of summer, but in spring it delays the start of growth by as much as three weeks. This shortening of the growing period may considerably reduce the total yield of forage.

Heavier grazing, however, leads to progressively greater and distinctly deleterious changes. This is most rapid and drastic on the richest type of prairie, since the dominant grasses there are the favorites of cattle. Cattle will hunt out big bluestem and eat that first. They also have a preference for Indian grass, switchgrass, and, at least until it grows hard and tough in midsummer, little bluestem. As the grass foliage becomes sparser, the root system is weakened and the whole plant grows less vigorously. The same fate soon overtakes Canada wild rye, needlegrass, and such typical prairie legumes as leadplant.

With shading and competition from the dominant grasses reduced or removed, a shift takes place in the relative abundance of the various plants. At first this involves only a little readjustment among the natives; but soon aliens begin to appear. In the drier situations these would be gramas and buffalo grass. Where it is a little moister, any bluegrass that has managed to insinuate itself increases markedly. With severe overuse, species that never appear in undisturbed prairie are able to invade.

The driest of Midwestern prairies are the least affected by severe grazing, since only native plants can endure the living conditions there. The wettest prairie bordering on sedge meadows is also relatively unchanged. But tallgrass prairie growing on moderately moist and fertile soil can be converted to an almost pure sward of Kentucky bluegrass by only a few years of constant heavy grazing.

Bluegrass needs no bare soil to get started and can infiltrate as soon as the tall native grasses become thin enough to let the light reach the soil between them. By the time the larger grasses disappear, a good sod of bluegrass may already have developed among and beneath them. If grazing is too drastic, even the bluegrass is reduced to a thin, patchy growth with scattered clumps of such rank weeds as thistle and ironweed — as unsightly a scene as it is deplorable, and all too familiar.

Just as light grazing benefits the prairie by reducing the mat-forming litter, so does an occasional fire. Burning undoubtedly improves the early growth of prairie grasses and gives a better "bite" for cattle in the spring; but whether fire is beneficial in the longer run has been much debated for a long time.

In recent years our understanding of what fire really does to vegetation is increasing rapidly, largely on the basis of controlled experiments. In a number of places tracts of land have been set aside specifically to study this question. These are located in places as diverse as the University of Wisconsin Arboretum at Madison, the Trelease Prairie at Urbana, Illinois, the University of Missouri's Prairie Research Station, and still others in Kansas and Oklahoma.

In all cases it is found that not only does growth start earlier in the spring following a fire, but the year's total production increases. Both the number and the size of flower stalks also increase — in both big and little bluestem by as much as five- to twentyfold. Even one year without burning appreciably reduces their growth. Repeated burning of the dead tops has little discernible effect on the roots, especially if it is done in early spring just before new growth appears.

The contrast between the native big bluestem and the alien Kentucky bluegrass is striking. Whereas grazing can convert native prairie to virtually pure bluegrass pasture, fire by itself can restore a good bluegrass pasture to bluestem prairie. Five years of annual burning was enough to accomplish this on one experimental field in Wisconsin.

Clearly resistance to fire, tolerance for a dry environment, and the ability to obtain water from deep in the soil, all have contributed heavily to the making of prairie vegetation.

In contrast to parts of the country that were originally forested, there is little land in the prairie region that has been cultivated for a time and then abandoned, for it is generally too valuable for farming to be left idle without some compelling reason. Land around growing cities may be left to its own devices while it is being held for speculation until urban expansion sends its value up, but it rarely remains undisturbed long enough to develop a stable cover. Except for areas that have been taken over for scientific study, about the only places that are left to revert to natural vegetation are stretches of roadways and railroad lines that have been discontinued.

Such places as there are show that once the disturbance is removed, prairie land will eventually be taken over by prairie plants. This was the case with one Nebraska roadway that was abandoned and fenced off in the late 1930's and subsequently never either burned or grazed. When it was found fifteen years later by an interested ecologist, it was covered with a typical 80 percent of big bluestem, the rest being mostly switchgrass and prairie dropseed. Often the return to prairie takes much longer than this, especially after a period of plow cultivation, when everything is removed except the crop plant and a respectably small number of weeds.

Restoration can be speeded up enormously by sowing seeds of the dominant prairie grasses. After the drought years of the 1930's attempts were made to repair some of the badly damaged pastures in this way. Seedlings of many perennial grasses are inconspicuously small at the end of their first growing season, but by the second year they begin to make a good showing. Mowing at this point suppresses the annual weedy types, to the great advantage of the developing perennials, and by the third year only a negligible number of annuals remain. Such a procedure saves many years of waiting while the process of natural succession goes its sometimes devious way.

Even without disturbance by man, beast, or fire, grasslands are periodically subject to disaster in the form of extreme drought. The appalling destruction of the 1930's was due in part to overoptimistic use of farming practices that were ill-suited to the climate; but in that catastrophe even undisturbed native vegetation suffered severely.

In the earlier years of the century the science of ecology had been developing rapidly. Among the people most active in this development were Dr. John H. Weaver and his colleagues and students at the University of Nebraska. In the course of their studies they had gathered a large amount of detailed information on the more normal state of the vegetation in the area that was to be worst hit in the thirties. With this background it was possible to follow very closely the changes that took place on the prairie as the drought progressed and later as the rains returned.

Throughout eastern North America the early thirties were unusually dry years, but the summer of 1934 brought the worst drought and heat that have ever been recorded on the prairie. That year the hot, dry, windy weather began in the spring, and by May the more shallowly rooted plants had already begun to dry up as water disappeared from the upper part of the soil. Plants with deeper roots were slower to suffer, and with them the first symptoms of lack of water appeared at progressively later times. Such plants as big bluestem and the extremely deep-rooted prairie rose were little affected. But in the heat and dryness plants that were not badly injured flowered and fruited prematurely. Goldenrods that ordinarily start blooming late in July were already showing yellow by mid-June.

As the summer progressed, with nothing but light showers that scarcely laid the dust, the soil dried to deeper and deeper levels. Although in many summers the top six inches or so of prairie soil become completely dry, that year as early as July there was no water at all to a depth of three feet. In August the dry zone fell to four feet, and even at six feet there was very little moisture available for even the most efficiently absorbing roots.

From mid-June to the end of July there was a heat wave

when the day's average temperature often exceeded 100° F. Midday peaks were usually ten degrees higher than that. Most of the time the air humidity stood below 25 percent, and on some afternoons it fell as low as 3 percent. With frequent strong winds and a completely cloudless sky for weeks on end, it was a cruelly searing time that descended on the grasslands. By August much of the vegetation had become crunchingly dry and as bleached of color as it usually is at the end of winter.

The next spring enough rain fell early in the season to moisten the ground to a depth of several feet. This revived such plants as had survived the first bad year and showed clearly the distinction between the living and the dead. In places that had been studied in detail before the onset of the drought, a third to a half of the plant cover was now found to be dead. Big bluestem disappeared over large areas, but it was able to hold out to an amazing extent in situations where its very deep roots could reach water that still remained at depths totally inaccessible to most other plants.

The drought continued for seven long years. In that time occasional rains wet the top part of the soil, but this did not last long. Below the moistened layer the earth was bone-dry to depths of six or eight feet, and always there was a thick dry zone between any surface moisture and the steadily falling level of deeply stored water. Through this barrier no roots could grow, and only those that had already grown below it could use the deep reserves.

Many plants held out for a time by virtue of their very long taproots; but as time went on even these became fewer and fewer. A few plants of prairie rose, leadplant, false boneset, and button snakeroot survived even to the end. Most of their dwarfed and meager shoots ceased to bloom, however, and for several years the changes that normally sweep over the prairie with the succession of flowers essentially disappeared.

With death and dwarfing so general, everywhere there were areas of bare, black soil, with neither foliage nor litter to break the force of the cruel sun. The large areas of exposed soil provided a heyday for some of the aggressive native weeds such as peppergrass, horseweed, and pigweeds; and much of what had been rich prairie came to look like an abused and weedy pasture.

On the western plains, conditions are always drier, and except in special situations there is none of the deeply stored water that supports the lush vegetation of the tallgrass prairie. Here the common plants have long been adapted to hot, dry summers and to soil that is practically always dry below the level of a few feet.

While the plains country suffered its own disasters in the drought years, some of the plains grasses were able to push eastward beyond their usual range as the taller prairie grasses fell away. Typical western species such as needlegrass and prairie dropseed, or the even more rampantly spreading western wheatgrass came in over large areas where the once abundant little bluestem was all but wiped out. Even buffalo grass, so characteristic of the western range, appeared in the nearby parts of the prairie country.

In 1941 the rains at last returned to normal, and gradually moisture penetrated into soil that had long been totally dry. It was several years before the moist layer worked down far enough to meet the deep ground water.

Although the remains of old grass clumps persisted and masses of roots were left underground, most of the plant matter was dead. Nevertheless, small nests of living tissue remained, and as these took up the newly available water, they began once more to grow. Sometimes it was two or three years before little tufts of new leaves showed that a spark of low-burning life had survived in a grass clump or a leadplant rhizome.

Within a few years the prairie once more had a complete green cover. But this was not at first the same as it had been. One of the first things to revive was big bluestem. All its surviving — although much reduced — clumps produced fast-growing runners. Often these could be traced by long lines of leafy tufts radiating out from the base of an old plant. With most of its strong competitors not yet fully revived, the especially wet year of 1944 brought bluestem such as had scarcely ever been seen before. Even on the drier uplands the foliage grew as large and rank as it normally does in the best moist places. Soon other species revived, and for several years the prairie produced a luxuriance of grasses of all kinds that was unequaled in human memory.

It was a little later that the brightly flowering plants had their surge of luxuriance. Large numbers of seedlings of all kinds developed from seeds that had been lying dormant for many years. One year there was more blue-eyed grass than is normally seen in several years combined.

In time the many members of the prairie community settled into a state of balance with each other. The western invaders slowly retreated in the face of competition and shading by the larger prairie grasses; but twelve years after the rains had returned it was estimated that the plains vegetation still occupied two-thirds of the land it had taken over.

Nothing like the drought of the thirties has been seen in the span of recorded history. It affected not only the prairies and plains, but also the oak forest region. In Iowa and eastern Nebraska, many trees were killed outright and many more were severely injured. If the drought had lasted a few centuries instead of a few years, it would have brought the kind of large vegetational shift that took place in the Xerothermic period of three or four thousand years ago.

When the westward-moving frontier of settlement came to the edge of the open prairie in Illinois, it came to a halt for a time. To European man, whose ways of living and getting a living had evolved in a forested country, life on a treeless expanse of grass looked difficult and forbidding. At first farmers were skeptical that such land would ever yield a decent livelihood. It took only a few trials, however, to discover the bountifulness of prairie soil.

In much of the Middle West one of the first chores in bringing the land under cultivation was drainage. Huge expanses of old lake beds, flat till plains, outwash sheets, and alluvial river bottoms were regularly covered with standing water from the first thaw until the start of really warm spring weather, or even for much of the summer. A chronic state of spring waterlogging may be acceptable to prairie grass, but a farmer must prepare the soil before he can plant his crops, and this cannot be done when the fields are a sea of mud. So from the beginning, miles of ditches have been dug and even more miles of underground drainage tiles laid. Even today deep, open ditches through the fields and along roadsides are a constant feature of Midwestern croplands and pastures. The tile drains that feed into the ditches, or in more rolling land empty into natural drainage channels, are less conspicuous and are likely to go unnoticed except when water is pouring out of their ends.

Widespread drainage as well as the prodigal use of underground water from wells soon brought about a general lowering of the water table. Possibly this as much as the ending of fires accounted for the rapid spread of trees in many places.

Besides the spontaneous spread of trees, settlement brought a flurry of tree planting. Old pictures of prairie towns that are now pleasantly green and tree-shaded show the first raw buildings standing stark and bare on the wide open, treeless grassland. As soon as a town existed, trees were set out along the new streets and around the houses. In the country, windbreaks were planted, both to protect the fields and farmyards and to serve as a source of wood.

Settlement of the prairie was virtually complete before it occurred to anyone to preserve specimens of the primeval grassland. For a long time now there has been no virgin prairie at all in Illinois, the Prairie State, and little enough of it anywhere else.

The most widespread surviving samples of wild prairie of any kind are found in a few old rural cemeteries and schoolyards that date from pioneer days and have never been "improved," and along the early railroads whose rights of way were fenced in before much of anything had happened to disturb them. These are mostly upland tracts, however. Many of the wet meadows and marshes that once were liberally strewn over the prairie have succumbed to man's passion for draining any place that is not either distinctly solid ground or distinctly navigable water.

The old cemeteries, "railroad prairies," and other small remnant bits may retain their primeval flora indefinitely. But if they are very small, and isolated from other tracts of prairie, they are likely to lose some of their species, for when some local disaster wipes out one group of plants or another, there is no nearby source of replacements. Consequently these places are not always representative of the original state of affairs.

More than thirty years ago ecologists had begun to urge the need of grassland areas for teaching and research purposes, comparable to the university forests. In at least two places this need has been met by converting pieces of old farmland into "synthetic grassland." At the University of Illinois, it was done by sowing the intended prairie with seeds of prairie grasses, which were mixed with a final planting of oats to serve as a nurse crop. After the grasses were well established, clumps of other prairie plants were brought in from the nearby countryside to encourage the development of a typical mixed vegetation.

This manmade prairie is divided into two parts by a road, which acts as a convenient firebreak between an area that is left undisturbed and another that is burned every few years. The burned part maintains a vigorous cover of big bluestem; but the unburned part has become strongly infiltrated with trees and shrubs. Since the whole area was once farmed, it is underlaid with a network of drainage tiles. It would be an interesting experiment, although probably not feasible, to plug up all the drains and see what effect that would have on both burned and unburned parts.

A similar process of seeding and transplanting has been used at the University of Wisconsin Arboretum. This place is designed as a study collection of complete living communities, including both plants and animals in all their variety. The twelve hundred acres of the arboretum cover a wide range of habitats. Whatever does not exist naturally is coaxed to develop by careful manipulation, yielding in the process much insight into causes and controls.

These activities give a special irony to the fact that John Muir once offered to donate a piece of choice Wisconsin prairie to the state, provided it would be left undisturbed in its primeval condition. But the legislature couldn't see the sense of such a thing and refused the gift — a source of profound regret to present-day scientists.

A few prairie tracts have been saved from destruction incidentally, like the Pipestone National Monument in southwesternmost Minnesota. This is a place where all the Plains Indians came for the special red stone from which they made bowls for their peace pipes. The first white man who left an account of the place was George Catlin, famous for his portraits of Indians. The prairie around the pipestone quarry looks the same today as it did when he painted a picture of it in the 1830's. It is a small sample, but one can view it with the satisfaction of knowing that it is purely natural and not the product of human manipulation.

11. Inland Waters: The Great Lakes and Their Shores

TO AN OUTLANDER familiar with the seacoast, the Great Lakes seem a paradoxical mixture of freshwater lake and ocean. All of them are much too wide to see across, and they are large enough to have true lunar tides; but the tides rise and fall in almost imperceptible magnitudes of two or three inches. They are subject to sudden violent storms that produce great, crashing waves; but much of the summer they lie calm as duckponds, and on a clear evening the water's surface is smooth as satin, scarcely rippled into wavelets that reflect the sunset sky — pale blue on one side, rosy gold on the other. The sandy wastes at the southern end of Lake Michigan are like those of Cape Cod or the Carolina coast, but the fresh breeze off the water contains no hint of salt. And although the rocky north shore of Lake Superior is sometimes likened to the coast of Maine, the rocks are dark and somber, with a deep wine-red or purple color. Any vignette of the Lakes must include the long, slim forms of ore boats, each trailing at intervals a thin plume of smoke from the funnel near its stern.

Geologically, the Lakes are rather unusual, for their watersheds are proportionately very small, the water itself covering about a third of the total drainage basin. Inflowing

On the north shore of Lake Superior the waves break directly on the rocky rim of the lake basin. The forest grows down close to the water here at the mouth of the Cascade River in Minnesota.

streams are short, measured at most in tens of miles from headwaters to mouth. They typically approach the lake in a sharp descent through a steep ravine. In the hard-rock rim of Lake Superior's north shore, the spate of early spring envelops the tumbling cascade of each little river in a cloud of pale golden foam. In the south, too, where streams flow over soft shale or thick layers of clay, the sides of lake-edge ravines are steep and sharply cut. These little valleys, filled with broadleaved woodland, have a characteristic look of muddy lushness.

The amount of water in the Lakes fluctuates widely, for their small catchment basins have only a limited area where heavy rain or melted snow can be stored in ponds and marshes or held in the form of ground water, to be released gradually and averaged out over the span of both time and distance. Water levels vary both from year to year and from season to season in the same year. Typically, water is highest in spring and early summer, after the year's snowmelt has reached the Lakes and before the rapid evaporation of the hot, dry summer has taken its toll. The year's low water is usually one or two feet lower and comes in winter and very early spring. These are vertical distances, and the shift of water's edge across a sloping shore is several times as great. The difference between the highest and the lowest levels ever observed in historical times is about six vertical feet.

The most spectacular change in water level is the sudden and unexpected phenomenon known as a seiche, or "tidal wave" — tidal in its magnitude, not in its cause. On the Lakes this results from an abrupt shift in the wind or an abrupt change in atmospheric pressure. When the wind has been blowing strongly down the length of one of the Lakes for several days, the water may literally pile up at the downwind end. The water in Lake Erie has been known to stand thirteen feet higher at one end than at the other. If at such a time the wind drops suddenly, the water surges back like so much water in a shifting dishpan, and the whole lakeful is set off into a series of gradually diminishing oscillations, each of which lasts for a matter of minutes or hours, depending on the geometry of things.

A seiche may also be set off by the sharp jump in atmospheric pressure that often accompanies squally winds and thunderstorms. Here is an incident recorded by one student of the beaches and dunes near Chicago. "The writer once camped on an Illinois beach, at least fifty feet inland from the upper reach of the waves. Shortly after the passing of a sudden storm, I was awakened by water spurting through the holes in the bottom of the tent; an astonished glance revealed that the shore line had suddenly moved fifty feet inland from the tent! Before we could drag the tents to dry land, the water receded, and later came back up partway and retreated in oscillations of many minutes duration, as expected in case of a seiche." Such are the hazards of scientific field work!

Any change in water level has a marked effect on the shoreline, widening or narrowing beaches, allowing the waves to wash against erodable bluffs or alternately protecting them by exposing a small, steep beach of gravel or cobbles, or perhaps a rampart of boulders from which all smaller stuff has been washed away.

Wherever they are found, beaches are constructed by the waves. As each wave rolls in toward shore, it stirs up the loose material on the lake bottom. Some of this it carries with it as it moves on, and when the wave finally crashes, both water and sediment are hurled up onto the shore. The receding backwash carries away only a part of the solid matter, and much of the heavier sand and gravel is left behind at the water's edge. The great waves of winter storms may shift large quantities of sand from place to place, robbing Peter to pay Paul and in the process remodeling the contours of a beach; but over the long run their work is chiefly constructive.

As soon as the waves subside and the sand begins to dry, the wind takes over on the beach. The loose particles are readily picked up by a stiff breeze, and where the strongest winds blow mostly onshore, sand may be carried inland and piled into dunes behind the beach.

Of all the specialties of the Lakes region, one of the most remarkable is the Lake Michigan dunes. There are large dunes in many places up the eastern side of this lake, but nowhere do they cover such a large area as at its southern end in Indiana. There the prevailingly onshore winds drive the sand-laden waves ceaselessly up onto the beach, and when the sand dries, blow it on inland to add to the growing dunes.

Since the slope of the land here is very slight, the sandy expanse has been much widened over the centuries by the falling water level. Near Gary, a drop of fifteen to twenty feet measured vertically has shifted the waterline lakeward by a distance of several miles. Behind the duneland of the present lake and separated from it in places by clearly recognizable morainal ridges, there are still other tracts of dunes that were formed during earlier and higher stages in the history of the lake.

All but the lakeward edge of this sand country has long since been covered with vegetation, but its contours are little changed since they were modeled by the waves and winds of long ago. Dunes, beach ridges, sand spits, windsweeps, all can easily be recognized under their cover of forest or prairie.

Going inland from the water's edge, there is a consistent pattern in the types of plants that one finds, and it is very easy to read this as a series of developmental stages, from dune grass through heath scrub and oak woods to beech-maple forest. This pattern caught the attention of botanists at the University of Chicago in the early days of the study of ecology, and it was there in the 1890's that Professor Henry Chandler Cowles brought into being the science of "dynamic ecology."

For convenience of discussion, any more or less sandy shore can be divided into three zones. Nearest the water is the lower beach, washed at intervals by even the moderate waves of summer storms. This is constantly changing, and no plants can get even a temporary foothold there. Above it is the middle beach, high enough to be undisturbed through the summer but still within reach of the large waves of the much more ferocious storms of winter. Except immediately after a rain, the surface here is very dry, and the annual plants that spring up every summer are typically succulent and drought resistant, like the sea rocket and seaside spurge. The upper beach lies above the highest reach of any waves and extends back to the edge of the dunes. It is commonly strewn with old driftwood and other debris, and this offers shelter from the wind for clumps of wormwood or evening primrose or a straggling growth of beach pea.

Clean beach sand is a remarkably homogeneous material. The finer particles of silt and clay have been winnowed out of it by wind and waves. Coarser gravel that may roll back and forth at the water's edge is too heavy to be thrown very far up on the beach by ordinary waves. Any humusy remains of dead plants or animals are rapidly reduced to an impalpable dark smudge. What is left consists of clean, fine particles of pure mineral sand, lovely under the bare feet of warm-weather walkers, but a very sterile substrate for the growth of living things.

As soon as the sand dries — a matter of minutes in the warm air of summer — it is subject to the action of the wind. A moderate breeze of about nine miles an hour is enough to move the average grain of sand and set it bouncing away over the beach. Above this threshold speed, the carrying power of the wind increases as the *cube* of its velocity. Consequently, even a short period of high wind can have a very large effect, as happens in winter gales or even on windy days of clear, brisk weather any time of year. At any given speed, the wind soon becomes "saturated" with all the sand it can carry, and anything that slows it will cause it to drop part of its load. Even while the wind holds steady, a piece of driftwood or a tuft of grass, even a change in the contour of the beach, can drag at the wind enough to touch off the start of a little pile of sand, an embryonic dune.

Very commonly the first persistent dropping of sand takes place just landward of the upper beach. This produces a long ridge parallel to the shore that is known as a foredune. Sometimes the foredune is broken into a series of separate mounds, but even then the general form of a beach-margin ridge can be recognized. So long as there is dry sand on the beach, the wind keeps moving it up the long gentle slope of the foredune. As the ridge continues to grow, some of the sand rolls on over its crest and down the steeper back slope. There it comes to rest at the angle of 32° that is characteristic for dry beach sand. The bottom edge of this steep slope is sharp as a knife, and the advancing front of a dune on the move forms a boundary that, for abruptness, has few parallels in nature.

On the upper beach and the adjoining dunes a patchy cover of plants leads a rigorous existence. Chief of the numerous hazards there is windblown sand. Not only is this strongly abrasive, but it is constantly being deposited around and over anything that protrudes above the surface. Any plant that is to survive in such a place must be able to grow fast enough to keep ahead of the rapidly accumulating sand. One of the few that can do this is beach grass — *Ammophila breviligulata,* to give it its formal name.

This plant is as closely attuned to life on the dunes as big bluestem is to the prairie. The secret of its success lies in its long, vigorous rhizomes, which keep pushing their growing buds up into the light as fast as the sand rises around them. At the same time they send an abundance of roots down to the depths where the sand is always damp. As the grass rises above the surface, more sand catches around its base, raising the surface. Above this the grass continues to grow, and so on, as long as the wind blows and there is sand left on the beach below. So the dune grows larger and larger.

Although the sting of windblown sand on the bare legs of summer shows that sand is on the move even then, most of the year's deposit is laid down by the high winds of winter. This seasonal difference is recorded in the grass plant's form. During winter, when growth has stopped, the plants become more or less completely buried, often entirely concealed beneath a smooth surface. When growth starts up again each spring, the new stems that form in the underground darkness become very long, and successive leaves may be separated by as much as four inches. As soon as the growing point reaches the light, however, stem growth is sharply checked, and leaves that form after that remain crowded together in tufts. The next winter sand once more accumulates around the old leaves, and the entire sequence repeats itself.

The hard, almost woody grass leaves persist long after their tissues are dead, and in any case there is always a recognizable joint on the stem where each leaf attaches. Consequently, on a carefully excavated plant it is possible to read off the history of a dune's growth. The testimony of the grass plants is that an actively growing dune rises at the rate of about a foot a year. If the deposition of sand is interrupted for any reason, the year's vegetation does not become buried during the winter but accumulates into a permanent strawlike mat on the surface. Hence, in early spring before growth starts, it is fairly easy to tell by their appearance which dunes are currently growing.

As soon as the waves no longer disturb it, any expanse of bare sand may be invaded by beach grass. Sometimes the new colony starts from a fragment of an old plant that has been broken off and blown or washed to a new place of rest. So long as the bud in its center is still alive, any tuft of leaves can continue to grow, producing both leafy stems and an abundance of new roots.

Not all new grass colonies can be accounted for in this way, however, and some of them must arise from seeds. Considering the relation between a small, sprouting grass seed and a dry surface swept by blowing sand, one may well wonder how seedlings ever manage to get a foothold. The question of just how they do it has been investigated on the spot in the Indiana Dunes.

Seeds of beach grass ripen during August, but they are not shed immediately from the seedheads. Instead, they are shaken out a few at a time as the tall seedstalks jiggle and bob in the wind. This goes on over a long span of time, and some seeds may remain in the head as late as March. Over the winter they blow about on the sand, and those that do not get carried into the water or eaten by birds come to rest and are buried in widely varying depths.

When spring comes and seeds begin to sprout in the warming sand, only those that chance to lie between one and four inches deep succeed in producing new plants. Nearer the surface, if they sprout at all, the sand around them dries out before their small roots are long enough to reach permanent moisture. If the seeds are too deep, the plantlets die of starvation before they reach light and air at the surface. Even some of those that make a successful start may die of desiccation if their tiny stems are not kept moist enough to induce the formation of stemroots; for the first taproot that pushes out of the seed, even after it has branched profusely, is not by itself adequate to provide the amount of water that the small shoot needs. To be sure of surviving, a plant must become well rooted during the cool, moist days of spring, for when warm weather sets in, the top few inches of sand dry out completely in a matter of minutes even after a heavy rain.

Once established, beach grass can grow both upward and outward indefinitely. In an established stand practically all increase is by the growth of existing plants, rather than by the addition of new ones. The outward spread can be traced by the long, straight rows of leafy tufts that arise from the rapidly growing runners at the advancing front of a colony.

As long as sand continues to accumulate around it, the beach grass thrives. It also continues to catch at the sand blowing through it, and the dune grows along with the grass. When eventually something interrupts the movement of sand from the beach below, however, the growth of the dune is sharply curtailed. It may be that a rising water level reduces the size of the beach, or that receding water or simply the steady action of breaking waves widens it so that it is only a matter of time until some impediment or roughness of surface touches off the start of a new foredune in front of the old one.

As soon as the grass no longer becomes buried every winter, its vigor begins to decline. This is not a matter of competition, for it happens before the grass has become very dense and while there are still very few other plants on the dune; nor does it seem to be a matter of smothering in its own thatch, as seems to be the case with big bluestem on the prairie. The most plausible suggestion, although it has not been experimentally tested, seems to be that the growing stem tips reach so close to the surface that they dry out in the loose, hot sand or the air above it. In any case, growth tapers off, fewer and fewer flower stalks are produced, and the dune grass goes into a general decline. For a long time, however, there is enough growth to impose a seasonal pattern on the rhizomes, so that the plants can be used as a sort of biochronometer.

As the surface of the dune takes on a degree of stability and living conditions become a little less extreme, other kinds of plants begain to appear. One of the most frequent is the sand reed grass (*Calamovilfa longifolia*). This plant grows taller than beach grass and has short, stocky rhizomes rather than long runners, giving it a clumpier habit of growth. It also lives longer by many decades than beach grass. Here also appears a special dune strain of the little bluestem that is so common on dry prairies. A host of other perennials appear among the grasses: goldenrod, horsemint, puccoon, and especially wormwood. Along with the perennials appear certain characteristic trees and shrubs, such as sand cherry, cottonwood, and some of the many kinds of willow. There are likely also to be tangles of poison ivy, grapevines, and catbriar.

Even as this cover of vegetation develops, sand continues to blow in around the plants, although much more slowly now. Some woody plants can tolerate this slower burial by virtue of their ability to form an abundance of new roots on their stems as the old roots become buried too deeply. Such plants also make good dune builders. The spreading sand cherry tends to produce low, broad mounds, while cottonwood makes tall, narrow dunes.

In places that are sufficiently sheltered from windblown sand but still fully exposed to the sun, a shrubby kind of heath may develop, with carpets of bearberry interspersed with hudsonia, shrubby cinquefoil, and quantities of the smaller false Solomon's seal (*Smilacina stellata*). Prickly pear cactus may come as a surprise to the dune walker, although it grows in hot, dry nooks in many places far removed from the desert. There may be a scattering of red cedars and common junipers, and where creeping juniper thrives, it forms such dense mats that nothing at all can grow beneath it.

Scattered over the duneland are sheltered hollows where moisture stands at the surface much of the time. Here seedlings of willow, cottonwood, and sometimes basswood find congenial living places. Where it is a little drier there may be pines, especially jack pines, which here are growing sixty miles south of their nearest brethren. Often there are black or white oaks.

As the trees grow larger, they gradually merge to make a woodland canopy. Pines last only one or two tree generations, but the oaks live on and come to dominate the forest. Where openings persist, or in places that are frequently burned, the woodland floor develops a bunchgrass type of prairie vegetation; but in the shade of the growing trees a typical dry woodland understory appears. Huckleberries and low growing blueberries form a dense mat of woody roots and rhizomes, and leaf litter begins to accumulate and slowly decay into humus.

It takes many tree generations to produce enough humus to make more than a black stain on the sandy surface; but as organic matter eventually increases, the nature of the understory changes. Carpets of such small plants as Canada mayflower and the common woodland sedge appear, and under the oaks arises a layer of the smaller sassafras, shadblow, and witch hazel.

In damper places, like slopes that face north or downwind, or the bottom of dune pockets, the soil is quite rich and fertile. Here grow basswood and red oaks, and with them small plants typical of moister woodland. Professor Cowles long ago suggested that these places mark the onset of another stage in the long vegetational succession which, given time enough, would spread over all the duneland.

In the seventy years since Cowles's pioneering work we have learned much about the interrelationships of plants and soil and also learned caution about accepting at face value many things that seem obviously true. Increasingly, ecologists have come to insist on a step-by-step, take-nothing-for-granted analysis of their evidence. In this framework some years ago Jerry Olson, then a graduate student at the University of Chicago, undertook a detailed study of the relation between developing soil in the dune region and the vegetation growing on it. His primary aims were technical and mathematical; but he was also interested in the question of whether or not the evidence really supports the classical theory of a "climatic climax," that is, that all change in a given region leads ultimately and inevitably to a single type of vegetation that is determined in the end by climate.

Olson worked along the Indiana shore, where the dunes have been expanding behind the receding water all through postglacial time. His first job was to find out how long the surface under each different kind of vegetation has been stable, that is, when each dune stopped growing. For this he used a variety of methods. For the youngest dunes he examined the buried stems of beach grass. For dunes old enough to bear trees, annual rings in the wood tell at least the ages of the trees, although the dune itself must be somewhat older. Finally, ages of the oldest dunes were determined by radiocarbon dating of old organic matter found in them.

Along the transect line laid out for this study, the youngest foredune ridge turned out to be eight months old. The one next behind it was about six years old. The third ridge bore pines up to eighteen years old. The first dunes old enough to have developed an oak wood with an under-carpet of blueberry have existed for six hundred to a thousand years. Highest and farthest from the shore, the soil and its vegetation have been developing for some twelve thousand years, ever since the postglacial beginnings of the lake.

Transformation of sand into soil starts with the burial of

the first pioneer generation of beach grass. The old grass tissues are very tough and remain as a binding framework for a long time; but eventually decay sets in and the plant remains are converted to humus. Once the constant burial by fresh sand has stopped and the grass thickens into a mat, it takes about four to six years for enough humus to accumulate to blacken the sand surface permanently. After that the old surface can be recognized even if it becomes buried again later.

Digging out old dunes shows that the roots of later generations of plants often exploit the somewhat richer black layers of former dune surfaces. In sand cherry dunes, for example, the great bulk of the cherry roots spread out in the succession of humified layers that have formed and then later become buried. Patches of especially lush dune vegetation of any kind are usually found to be rooted in a pocket of old organic matter that they have come upon by some happy chance.

While humus is slowly accumulating, other changes take place in the sand. Iron that was originally present in a gray, chemically reduced form becomes oxidized, staining the sand particles a yellowish orange. This change works its way slowly down from the surface. So does the leaching or washing out of limy compounds. As lime is lost, the sandy soil becomes strongly acid. In a thousand years most of the soluble lime has been removed to a depth of about six feet in the dune region. This is well below the reach of the great mass of roots of the plants that grow here.

As these changes go on, the sand acquires minute amounts of finer silt and clay. Some of this is probably what is winnowed out of the sand on the beach upwind. The rest is formed by chemical breakdown of minerals of the sand itself. Even after long spans of time, the amount of fine material is still extremely small, and it would hardly be noticed if it were not entirely lacking at the beginning.

Slowly also there is an accumulation of the various mineral salts that plants require for their growth. Nitrogen is added, probably by the activities of those remarkable bacteria that can take it from the air and convert it to their own substance. Phosphorus is freed by slow weathering of sand minerals, although it never becomes adequate to support really lush vegetation. In the Indiana Dunes sulfur comes in plentiful amounts in the rain, which washes it out of the air moving in from the industrial regions upwind to the west.

All these changes and additions go on extremely slowly, and soil fertility is very low for a long time. Anything that can live on such raw stuff has a quite remarkable ability to make efficient use of sparse resources, especially nitrogen. The first woody plants that appear are often pale and yellowish. Even when they look normal, a small dose of fertilizer applied experimentally has a spectacularly stimulating effect.

The oaks and the blueberries and other plants that grow with them on these meager sands are undemanding, taking rather little from the soil; but they also return rather little to the soil when they die and decay. Their leaves disintegrate only slowly, becoming highly acid in the process. This further acidifies the sand and increases the solubility of such minerals as there are, so that they rapidly wash away through the highly porous sand. The net result is that after the first thousand years, vegetation and soil come into a state of equilibrium, where as much is lost as is gained. Soil dating back twelve thousand years is not perceptibly different from the soil of dunes only one thousand years old, and it is hard to see any possibility of change in the system that could lead on to a soil capable of supporting the theoretical "climatic climax" forest for this region.

The few places in the dunes that harbor a richer forest of basswood, sugar maple, and red oak have a deep, fertile soil that seems also to be based on old dune sand. These are the spots that Professor Cowles interpreted as outposts of a regional climax forest. But careful inspection shows that in all such places, the soil is consistently moister, being especially favored from the start by shelter from the midday sun or from prevailing winds. There is no indication that the maples and basswoods are spreading or that they will ever replace the adjoining dry forests of oak.

Even after it is firmly established, the forest does not necessarily have a permanent hold on the dunes. Anything that causes a break in the cover and exposes bare soil can set the sands to moving again. Denudation by fire, erosion by waves during high water stages or exceptionally violent storms, even the wearing of a new path to the beach can do it. Once the wind gets at the soil, it does not take long for it to excavate a windsweep or "blowout." The result is likely to be rather spectacular, not only for the long, cleanly gouged channel, but perhaps more for the advancing sandhill at its downwind end. This is the kind of dune that migrates inland over long established marsh meadows or forest, burying live trees and exhuming the long-buried skeletons of old ones. In its wake one can see the holding power of a dense meshwork of grass roots and the ability of the pioneer dune trees to produce roots above roots above roots as they are progressively buried.

Once a windsweep starts, it usually keeps growing until it reaches a depth where the sand is permanently damp and hence resistant to blowing. This leaves a long, flat-bottomed, steep-sided gully in the middle of a tonguelike mass of moving sand. A windsweep may remain active for a long time, but its days are ultimately numbered; for eventually a foredune will rise across its mouth, intercepting both sand and wind. Windsweeps have been forming and stabilizing for hundreds and thousands of years, and the contours of many old ones can be readily recognized, even under a cover of heavy forest.

In spite of the dramatic migration of active dunes, the Great Lakes region has no permanently wandering dunes such as one finds in a great sandy desert like the Sahara. In this climate all dunes eventually become covered with forest or prairie. They are, however, constantly replaced by new ones blown up from the beach below.

Northward from Indiana on both sides of Lake Michigan,

dune building is relatively slower and erosion of the shores much greater. Up the eastern side of the lake, especially toward the north, the dunes stand upon level-topped bluffs that may themselves rise more than three hundred feet above the lake surface. Best known of all these is the Sleeping Bear, near Glen Haven, Michigan.

When the Bear was first seen from the lake by white men, it was a huge pile rising more than 135 feet above the high plateau on whose edge it was perched. Seen from a lake steamer in the 1890's, with its double-humped silhouette and a dark cover of forest, it looked very much like an old bear sleeping on her side, not far from her drowned cubs, the Manitou Islands.

The first detailed ecological reconnaissance of the Bear seems to have been made in 1928 by a class from the University of Michigan's biological station at Douglas Lake, fondly known to many generations of students as the "Bug Camp." At that time it was evident that the big dune had been stable for a long time, for its northwestern face was covered with a heavy forest of maple, hemlock, and white cedar, which then reached to the very edge of the bluff. The other sides were also fairly well covered, although the vegetation there resembled earlier stages in the development of dunes. On the northeastern side a trail through the brush had been worn down to expose loose sand, and this showed signs of turning into a windsweep. The southwestern side was a slope of bare sand, sparsely strewn with the bleached and polished skeletons of trees, and into this the wind was cutting sharply.

The base of this eroding slope stood at the very edge of a high bluff whose sheer face showed that it was being undercut by waves eating into its base. From this it seems likely that erosion of this side of the Bear had first been touched off by action of the lake far below.

However and whenever erosion began, the rate of loss from the Bear has been impressive. In 1932, as part of a general survey of the Lakes region, the U.S. Coast and Geodetic Survey planted a base mark on the very crest of the great dune. This was the standard marker consisting of a metal disc set in the top of a twelve-foot pipe, the whole anchored in a large block of concrete. The makings of this apparatus were hauled by horses up the side of the dune exposed directly to the prevailing wind. The track that they made soon became a well-marked footpath. A year later the pipe, which had been completely buried, protruded a clear eight feet above the sand, and the year after that it was nowhere to be found. Local opinion was that it must have fallen into the lake.

At about that time the dune really began to blow away, and sand from the windward side and especially from the top moved out over the adjacent fields. In the course of the years since then, what once had been a long, shorewise ridge rising to a sharp pinnacle-like crest was first gouged into a horseshoe form and then cut completely through to form two mounds. The eroding face has cut back to the crest of the dune and beyond, so that the highest remaining point has shifted eastward and in the process become lower.

At the same time that the marker disc was set on the dune's peak, a similar marker was set into a boulder that stood on the bluff top due north of the marker on the Bear. Since then not only the dune but also the rim of the bluff on which it stands has been retreating landward. Measurements taken from the boulder marker show that in thirty years the bluff's edge receded more than fifty-three feet. From this it seems clear that, although thousands of sightseers tramping up the trails have greatly accelerated the destruction of this old landmark, waves of the lake below have long been working in the same direction. With its very underpinnings being cut away, its years were no doubt numbered in any case.

The Sleeping Bear is just one special, well-known instance of what is going on along this entire stretch of shore. The waterline here is particularly susceptible to erosion, partly because the wind is prevailingly onshore, but chiefly because of the contour of the land.

The edge of the lake basin here follows the border of the

last retreating ice sheet. This part of the ice margin was very irregular and was fringed with many sharp lobes. As the ice lay finally melting, the massive Manistee moraine accumulated along its edge, with thick fingerlike protrusions reaching into all the narrow gaps between successive ice lobes. The huge moraine forms a belt of high country that parallels this shore, and its interlobate fingers lie almost at right angles to both the general trend of the moraine and the line of the water's edge.

When the lake first began to form between the melting ice and the high moraine, the shoreline must have been much more irregular than it is now, for in the millennia since then the waves have been cutting back the protruding points and filling the intervening bays with what they have cut from the headlands, meanwhile building beaches all along the shore. The onshore winds are often strong enough to blow the beach sand right up the face of the bluffs and onto the top of the moraine, where it comes to rest in the form of dunes. Although nothing else seems ever to have been as spectacular as the one-time Sleeping Bear, the same processes that made it are still building other dunes all up and down the shore.

All these changes went on also during earlier and higher stages of the evolving lake. Behind the present shore one can recognize old dunes and beaches, abandoned wave-cut cliffs, and shallow bays that were more or less filled in by the waves of other times. Old bays that were fairly shallow now stand high and dry. Others are wet and marshy or are filled with typical northern bogs. The deepest of them still hold water in the form of landlocked lakes such as Crystal Lake, Glen Lake, and many others. Some of the fossil bays make perfectly symmetrical amphitheaters, like the one that encloses the hamlet of Arcadia. Not far from there, at Betsie Bay, the highway circles around the beach at the head of just such a bay, right under the old bluffs, then climbs high up onto the moraine top, from which one can see miles and miles of other bluffs, bays, and headlands. This is an exhilaratingly spacious country, beloved of many generations of vacationers who have the good fortune to know it.

The western side of Lake Michigan, in Wisconsin, has a similar history of erosion and wave action, but there the effect is less severe and it lacks the after-work of strongly onshore winds. Consequently, there are some fine beaches and much water's edge bluff, but relatively little in the way of dunes. A special feature here is the series of low, sandy ridges and alternating swales that parallel the shore in many places. One such place at Bailey's Harbor on the Door Peninsula has been set aside as a nature preserve, the Ridges Sanctuary. Other ridges appear at Point Beach State Park, and in fact anywhere that offshore sandbars have escaped destruction by waves and wind as the water level has continued to fall. In earlier times they were common also from the Indiana Dunes westward around the end of the lake through Chicago; but of course all natural features in this region have long since disappeared.

The ultimate origin of such sand ridges lies offshore at the point where waves break as they move in over a gradually shoaling bottom. The process seems to be as follows. In the waves traveling over the lake, water is quite literally in a state of rotation. The surface ridges that we call waves are the bulging tops of rollers whose lower parts are hidden under water. When a wave rolls into water of a certain shallowness, its lower side touches the lake floor, which drags at the bottom of the wave enough to slow it down a little. Since the top of the wave continues to move freely, it outstrips the water beneath and pitches forward in the form of a breaker.

All this time the bottom side of the rolling wave is moving lakeward. As each wave touches the bottom, it scours a trough, with a small ridge just behind it. The reduced wave then moves on toward shore, and at a certain lesser depth the whole process is repeated, forming a second ridge and trough. Usually this happens a third time before the wave finally washes up onto the beach and comes to an end.

While the level of the water remains constant, the three ridges seem to remain constant also. When the water is calm and clear, the three offshore ridges, sometimes exposed as bars, can be followed for long distances by an observer in a low-flying plane. With no tides rising and falling daily, the ridges may persist for a long time.

When for any reason the water level falls, one or more of the ridges may be left standing above water. If waves breaking on them then deposit more sand than they wash away, the ridges may become massive enough to survive the force of even severe storms that come later. Once safely out of reach of the waves, they are preserved indefinitely. Meanwhile, more ridges will form at the new and lower water level, and so on more or less as long as the lake lasts.

The vegetation that develops on such shoreline ridges is much like that of other sandy places in the vicinity, whatever that may be, whether oak forest or pine or, farther north, balsam fir.

What grows in the swales or sloughs depends on how wet they are and how long they have been developing. At first they usually contain standing water, and if it is shallow enough to become quite warm, it is likely to fill up with rushes and cattails. These are replaced in time with a dense thicket of alder. Deeper and colder water, especially northward, may turn into a typical cold bog. One of the charms of the Ridges Sanctuary is the profusion of orchids among the many bog plants there.

Eventually the swales become drier as organic matter accumulates and mineral sediment washes in. Some of them have become drained by a general lowering of the water table. Better aeration encourages the growth of trees, and in time the swale fills with damp woods. The soft, soggy muck around their roots is not very stable, and trees are easily blown over, sometimes making wildly impenetrable thickets. In the recurrent times of high water, the soil may again become totally waterlogged, and if this condition persists, many trees are killed. But temporary setbacks notwithstanding, the long trend is toward better drainage, and the forest in time becomes permanent.

Long stretches of the shores of all the Lakes have neither dunes nor ridges but are lined with steep bluffs or cliffs that drop off sharply, sometimes directly into the water, more often onto a narrow, steeply sloping beach of some sort. Some of the cliffs are cut into rock, as along the south shore of Lake Superior near Munising and the north shore above Duluth, and also around some of the islands in the western end of Lake Erie. Most of the bluffs, however, are carved in glacial till or the clay of postglacial lake bottoms, or in soft shale that erodes almost as easily as clay. Undercutting by the waves will in time cause any surface, solid or loose, to give way; and large blocks of earth and rock that have broken off and slumped or slid partway down the bluff are a familiar sight along these shores.

Lakeshore bluffs are subject also to washing and wasting by rainfall and runoff water. When the clay or fine silt of which many of them are composed becomes thoroughly saturated, it turns into a semiliquid mass of mud that can suddenly start to flow, sometimes quite rapidly.

Erosion of the bluffs is revealed by the tilt of trees and the protruding lip of soil on the top, held in place only by a binding mass of roots. In any season a generation of annual plants may grow up on freshly exposed soil, but most of the vegetation on the bluff face is likely to have started life on the level top and be in the process of sliding down into the water.

Where eroding shores lie adjacent to valuable real estate, large numbers of stone groins or piers have been built out at right angles to the shore to trap the sand being carried by longshore currents. The resulting triangular beaches show very clearly which way the currents are moving. The beaches are useful in themselves for recreational purposes; but they also serve to protect the erodable cliffs above from the destructive force of the waves.

All around the Lakes the influence of man is strong.

Large expanses of the original shore have been completely obliterated by the spread of cities, industry, and resort developments. In some places land has been built out into the lake for parks, airports, or general urban expansion. Elsewhere, farm fields reach right to the edge of the bluffs.

The effluent from all this human activity is increasing rapidly and has already become so enormous that the lake water itself has been drastically changed — a story beyond the scope of this account.

In spite of the changes, the Lakes still support a substantial commercial fishery, and they provide all kinds of aquatic recreation for large numbers of people, a significant factor, considering that some ten million people live within a hundred miles of the Indiana Dunes alone.

The Lakes are also an enormously important means of transportation. Even before the St. Lawrence Seaway opened, the tonnage that passed through the locks at Sault Ste. Marie, far up the Lakes, although it consists largely of grain and iron ore, long exceeded the traffic passing through the canals at Suez and Panama combined.

All this is in spite of the fact that navigation is closed for about half the year, generally from November to April, but depending on the weather and differing from lake to lake. None of the Lakes freezes over completely, and modern icebreakers are highly efficient; but much ice forms along the shore, sometimes extending out for several miles, and many large bays entirely freeze over. Large masses of floating ice are blown around by the wind, and often the ice piles up alongshore in a colossal jumble of huge blocks and cakes. This sort of thing can do severe damage when it grinds against shore installations such as breakwaters and boathouses. Just the process of freezing and thawing can damage small structures, and boats and small landing docks are usually taken out of the water for the winter.

In view of the enormous value of the Great Lakes, which mean so many things to so many people, the problems of pollution and changing water levels are currently receiving a great deal of public attention. It may be that the present few years may see the turning point away from a progressive worsening of pollution.

Fluctuating water levels are another matter. Lake Superior and Lake Ontario are already controlled to a large extent by locks on their outlet rivers; but the others suffer a natural regime of alternate feast and famine. With high water come widespread flooding and sharply accelerated shore erosion. Beaches are narrowed, and some of them disappear altogether. Small boats become landlocked because they cannot get under the low, fixed bridges between sheltered marinas and the open lake. In low-water years the stony or muddy lake bottom is exposed beyond the beach, and docks and piers are left stranded some distance from the water's edge. This not only brings disaster to owners of shore resorts as well as to hordes of wildlife, but the economic impact on hydroelectric power and water-

The old lighthouse at Marblehead in the western part of Lake Erie stands on an outcrop of Niagaran limestone across the bay from Sandusky, Ohio.

borne freight ramifies widely through the whole regional economy.

The open water of the Lakes offers no problems to navigation. It is the approaches to harbor facilities and the shallow connecting rivers that are the bottlenecks. In these places channels are kept dredged to a minimum depth that takes care of ordinary low water, and shippers calculate their lading accordingly. But in the once-in-a-century extreme, when there is not enough water to fill the channels, cargoes must be reduced so that ships will ride high enough to clear the shallowest passages on their routes. This, of course, raises costs.

Many schemes have been put forth over the years for stabilizing the water levels in the Lakes. One of the most audacious is to divert some of the Canadian rivers that flow northward into James Bay. The water would be impounded near the bay, then sent south through the Harricanaw River, reversing its flow, then through a cut in the height of land into the Ottawa River, through another cut into Lake Nipissing, and finally into Georgian Bay. Whether this rather horrifying piece of engineering ever materializes remains to be seen. But the importance of the Lakes and their commerce is such that it or something like it may yet come to pass.

12. Lesser Waters: Lake, Marsh, and Swampland

STREWN OVER the varied surface of the Middle West are thousands upon thousands of lesser lakes, ponds, swamps, marshes, and other assorted kinds of wet places. Besides the well defined lakes and rivers, there are large areas that are neither deep enough to be considered bodies of water nor dry enough to qualify as solid land. Some of these are the morainal sloughs ("sloos") and marshes of the prairie country. Others lie in backwaters of rivers or in shallow lagoons cut off by bars at the river mouths. Still others are the dying remnants of postglacial lakes and waterways, like the floor of Lake Wisconsin and the fringes of Sandusky Bay at the western end of Lake Erie.

All through the Middle West the rougher of the end moraines contain literally uncounted ponds and lakes. Minnesota alone is said to have far more than the ten thousand claimed in its patriotic slogan. Wisconsin has thousands more, and there are distinct "lake regions" in northeastern Indiana and in northwestern Iowa and nearby South Dakota.

Away from the rougher glacial country, natural lakes and ponds are usually related in some way to sizable rivers. Sometimes they occupy low spots left during the construction of a river floodplain, or they may be shallow impoundments caught behind the natural levees that parallel many stretches of river channel. Many lakes have arisen as oxbows, those deeply looping segments of a meandering stream that were cut off from a river during some time of high water.

Special paradoxes are the swampy areas that lie in what once were large drainage channels but now stand on continental watersheds. Such are the Horicon Marsh in Wisconsin, the Fox-DesPlaines watershed in Illinois, and the St. Croix–Brule region in northern Wisconsin. The enormous muskeg that fills part of the old bed of Lake Agassiz in northern Minnesota lies close to, although not really on, the "height of land" between Hudson Bay and the St. Lawrence drainage basin.

Even now, after so many generations of ditching and draining, there are times after periods of heavy rain or during the spring runoff when all man's drainage works become saturated. Then something of the old swampiness and the profusion of ponds and lakes reappear for a while, and a trip in a low-flying plane gives a glimpse of what a place like the Iowa prairie was like in its primeval condition — providing one's mind's eye is capable of erasing the pattern of roads and the rectangularities of field boundaries.

Among so many lakes spread over an area as large as the American midland the variations are almost endless. To bring some measure of order into the profusion, limnologists, the students of lakes, classify them into two basic types, which they call oligotrophic and eutrophic. The names refer to how well the lakes support life.*

To pack many technicalities into a nutshell, a typical oligotrophic lake is deep and cold. The water is soft, with little dissolved matter, and often quite acid. Sediment accumulates very slowly on the lake bottom and is always rather scanty. Such lakes are not very productive of either

* The words come from the Greek. *Oligo-* means little or small; *eu-* means truly or well. *Trophic* refers to nutrition.

plant or animal life, although they are the characteristic habitat of trout. The rockbound lakes of far northeastern Minnesota and the long, narrow ones east of Grand Traverse Bay in Michigan are prime examples of oligotrophic lakes.

The typical eutrophic lake is shallower and its upper part becomes warmed to a considerable depth in the course of the summer. The water is hard, since it contains a good deal of dissolved mineral matter, especially lime. Such a lake can support an abundance of algae,* and the shallow margins and loose bottom sediments offer both roothold for larger plants and burrowing and skulking places for creatures of many kinds. This makes a poor living for trout but a fine home for perch, pike, or bass. The great bulk of morainal lakes and essentially all riverside lakes are at least moderately eutrophic.

Given time, sometimes on a geological scale, oligotrophic lakes in the natural course of events tend to become eutrophic. As one might expect, there is a wide spectrum of intermediate stages. Although the process of eutrophication is essentially normal, it is enormously speeded up by almost any kind of pollution, and in recent years "eutrophication" has become a familiar word in the public press. Perhaps a better word here would be "super-eutrophication."

Many hard-water lakes in the Middle West have an accumulation of the limy sediment known as marl. This is a product of the stonewort alga, *Chara.* These remarkable plants incorporate so much lime into their tissues that they have a rough, brittle texture, and where they are abundant they may deposit as much as half a ton of marl per acre per year. Other water plants such as the water weed Elodea may also precipitate lime, but no others are such active marl formers as the stonewort.

In lakes where the water is very clear, submerged plants

* Pronounced "al-jee." This is the plural form. One of them is an "al-ga."

may grow at great depths. The record seems to be held by the appropriately named Clear Lake in Vilas County, Wisconsin, where two kinds of aquatic moss have been found growing under more than sixty feet of water. The critical factor is the penetration of light, since plants can grow only where light is strong enough to allow their production by photosynthesis to be greater than their consumption by respiration. In muddy water this may be a matter of only a few inches below the surface.

Sometimes the water is clouded not by mineral mud but by extremely fine organic matter. In many northern lakes the water is transparent enough but is stained brown from peat or other decaying vegetation. Such lakes are usually acid and quite unproductive.

A number of experiments have been done to see what adding lime would do to such acid lakes. One of the first trials was made on Cather Lake, a small one in northwestern Wisconsin. The result was striking, as the brown cloudiness disappeared entirely, leaving the water clear and colorless.

This result prompted a full-scale study of a pair of interconnected lakes known as Peter and Paul. Together these had an hourglass shape, making it easy to block off the narrow channel between them with an earthen dam. Then Lake Peter was limed, leaving Paul to serve as an untreated "control." Soon after the lime had been added the water began to clear, and two years later the same light intensity measured at a depth of eight feet in Lake Paul was found at more than twenty-one feet in Lake Peter. With this came an increase in the plant life, as well as in the density of microscopic plankton animals. These changes would in turn increase the population of animals right on up the food chain.

Where the lake edge is shallow and muddy, it supports a profusion of plants. This is the typical home of the widespread and familiar cattails, often intermixed with reeds and rushes. In spaces among the tall leaves the dark surface of the water is decked in season with three-parted white flowers of arrowhead or bright purple-blue spikes of pickerelweed, frequented by many a dragonfly. Here too grow the spectacular giant reedgrass (*Phragmites*) and, in more northern regions, wild rice.

Wild rice is a truly aquatic plant, for its seeds germinate under water and the seedlings live totally submerged until they are tall enough to break the surface. The plants ultimately reach the impressive height of seven or eight feet. They produce an abundance of seeds that are still commonly harvested by bending the tall heads down and knocking the seeds off into a canoe. Most of the wild crop is gathered by Indians on their own reservations. After passing through the hands of various middlemen and being processed and packaged, this product of the northern lakes and marshes ends up as an expensive delicacy on the tables of city dwellers.

In a good year millions of pounds are harvested, since the plant grows abundantly over a wide area. Its abundance is reflected in the fact that Minnesota alone has 76 Rice Lakes — along with 99 Long Lakes, 91 Mud Lakes, and a scattering of Rat (for Muskrat) Lakes.

In recent years a number of "paddies" (a strange word for their setting!) have been developed in northern Minnesota, where wild rice is being grown as a commercial crop. Although plants in the wild state may grow in water as deep as five feet, they also do very well in as little as six inches. Since the farming operation depends on the use of machinery, the water is kept as shallow as possible, and the paddies are allowed to dry down in late summer so that combine harvesters can operate.

Such plantations have yielded almost ten times as much rice per acre as the wild stands. If strains can be developed that will not shatter in the head before they are harvested, the yield might increase another tenfold. Should this venture prove to be successful, although it may not do much for the retail price, it would certainly increase the market supply of this faintly grassy-tasting but delectable food.

The more northern lakes often have remarkably smooth

and even outlines. Those that originated as glacial kettle-holes have always been this way; but others began life with quite irregular outlines and have become evened out over the centuries by the action of winter ice.

When water first freezes, it expands forcibly; but after that as it grows progressively colder it behaves like other substances and begins to contract. Cracks that appear and gaps that develop between ice and shoreline become filled with still unfrozen water which then freezes and adds to the bulk of ice. As the temperature drops still lower, the whole mass of ice continues to contract, often with resounding booms that can be unnerving to the uninitiated.

When eventually the temperature rises and the now over-sized mass of ice begins to expand, its edges exert a powerful push on the shore of the lake. Loose soil, especially the sand or gravel of glacial deposits, is bulldozed up into a low but abrupt ridge. After the thaw, this remains as a raised rim just above the water's edge. Over the years this process knocks away any protruding points or peninsulas and constructs barriers across the mouths of bays and inlets, rounding and smoothing the contours of the shoreline.

A cover of ice over a lake markedly alters the conditions of life in the water beneath it. With no waves ruffling and stirring the surface, there is no mixing of air into the water, and the supply of dissolved oxygen sharply decreases. The ice also screens out much of the light, although if it forms quickly enough on a still, very cold night, it may be quite smooth and transparent. Then microscopic green plants may be able to produce enough oxygen to make an important difference to the winterbound animals of the lake. Snow on top of the ice still further reduces the amount of light that gets through to the water.

In a deep oligotrophic lake the sheer volume of water may contain enough oxygen to meet the needs of the relatively few creatures that live there; but in the more fertile shallow lakes many fish suffocate during a severe winter, and when the ice finally goes out in spring, their limp corpses appear bobbing in the shallows and drift into windrows along the water's edge.

Toward the western edge of the prairie region, especially on the high, irregular moraines of the Coteau des Prairies, there still remain thousands of the small ponds known as prairie potholes. Some of these dry up by the end of summer, and many more disappear in times of protracted drought. Those that persist tend to be alkaline or brackish and support a growth of such typical brackish water plants as the ditch grass, *Ruppia,* and its fragile-looking relative called naiad.

Some of these ponds are highly productive of fish; but it is the bird life that gives them their special, lively charm. The pothole country, extending from the eastern Dakotas into nearby Canada, is said to be the hatching place of fully half of all the ducks in North America. In spring, when rainfall is abundant and all the birds are breeding, the potholes teem with waterfowl, and even a water-filled roadside ditch has its pair or two of coots or shovelers or twirling phalaropes.

It is the marshy shallows full of reeds and cattails that are the great producers of wildlife. Be it birds, fish, insects, furbearers such as mink and muskrat, or the frogs, turtles, snakes, and their cold-blooded kin known familiarly to naturalists as "herps," the marsh offers home and a living for countless creatures. The diversity of both animals and plants produces a richly complex food web, from predator through prey to the seeds, tubers, and rhizomes on which herbivores feed, and on down to the fungi and bacteria that live on the remains of them all.

The life of a well-developed marsh rests on a foundation of silt, peat, and muck. This offers a potentially rich foundation also for man's crops, and large areas of drained marsh have joined the one-time prairie in the total of profitably cultivated land. Many drainage ventures, however, have turned out to be something less than shining successes. When the great drought of the 1930's put many farms out of business, drained marshland was abandoned along with the rest. In its dry state, the thick peat that underlies so many northern marshes was burned out, often right down to bare mineral soil. When the rains returned in the forties, much land that had always been marginal at best had by then become totally useless. Large areas of this passed into public ownership, either forfeited for nonpayment of taxes or purchased outright by various public agencies. Both Wisconsin and Minnesota have extensive public lands that were acquired in this way, and some of them are now managed for the benefit of wildlife.

Some of these wildlife preserves have had complicated histories. Horicon Marsh, on the upper Rock River northwest of Milwaukee, was first manipulated in 1846, when one William Larabee dammed it to make a lake for his Tamarack Lodge resort. Then in 1868, at a time when hunting clubs were fashionable and hunting for the market was a common practice, the dam was removed to restore the marsh. Later, as farming spread into the area, the marsh was drained for cultivation; but as so often happened, the farms were never very successful. Moreover, in dry seasons the peaty soil often caught fire. Eventually it became clear that the best use of the land was its original one of providing a marshy home for wild things, and in 1940 public land acquisition was begun for that purpose.

At present the marsh covers an area about fourteen miles long and three to four miles wide, in which the water level is controlled by dikes. About twenty-one thousand acres of this is included in the Horicon National Wildlife Refuge. The refuge also owns some of the adjacent dry land, which it farms on shares with local residents. The refuge's share of the harvest is left standing in the fields for the benefit of migrating birds.

A similar story could be told about Thief Lake and Mud Lake, which lie in shallow depressions on the floor of Glacial Lake Agassiz in northwestern Minnesota. These are now included in the Lake Agassiz National Wildlife Refuge.

Paul Errington, a longtime and loving student of the prairie marshes, wrote "I saw Mud Lake as a desolation of burnt or smoldering peat grown up to little except farm weeds" at the end of the great drought. Eventually the rains came back; and when these tracts came into public ownership the lake outlets were dammed. Soon the old marshes and shallow lakes returned to essentially their original wild state, although here, too, the water level is now controlled. The wide view over this flat expanse of marsh and pond from a twenty-foot tower in the month of May is one to expand the heart of any bird-lover!

Even without artificial draining, the shallow lakes and marshes of prairie country often dry up in times of drought. Errington describes the floor of Lake Albert, near Brookings, South Dakota, as covered with hay and barley fields in 1934. By 1941 it had once more become a marsh, and in 1946 it was again a full-fledged lake with an expanse of open water. All this was without benefit of any human intervention.

In a number of places where formerly drained land has been reflooded, the new marshland supports a large waterfowl population for the first few years; but after a while the birds lose interest in it, and its productiveness (of birds, that is) seems to taper off. If the water level is then lowered for a season or two so that the marsh bottom is just barely exposed, the marsh vegetation is rejuvenated (from the point of view of waterfowl). The most important effect of this temporary "drawdown" seems to be to increase the amount of emergent vegetation — plants that are rooted in the bottom but whose upper parts protrude above the water. The temporary exposure to the air improves the fertility of the soil and permits the start of a new generation of seedlings. There is little change in the array of species, just a reinvigoration of the marsh plants on which waterbirds depend for food and cover.

In unregulated wild marshes the same kind of reinvigoration follows the periodic "drydowns" that come with the intermittent droughts of the variable Midwestern climate. Here is another instance, like recurrent fires, when what seem like disasters are in fact regular features of the natural scheme of things.

Just at the edge of a marsh or lake there is often a very slightly higher zone that is not so much flooded as perpetually waterlogged. This is covered with a grassy-looking growth that consists chiefly of sedges, distinguishable from true grasses by their triangular stems.

The usual sedge meadow is strongly dominated by two kinds of plant: the bluejoint grass and a large tussock-forming sedge by the name of *Carex stricta.* There are also a number of lesser sedges, and through the seasons the meadow is brightened with a succession of boldly colorful flowers, from marsh marigold and iris to the strong yellow of beggar ticks and assorted asters and goldenrods.

In spring, when it is flooded with standing water, a sedge meadow teems with breeding wildlife. It is loud with the calls of woodfrogs, spring peepers, and all their froggy relatives. Here carp and black bullheads spawn, and in the tussocks mice and shrews rear their families. The meadow is home for many early-nesting birds — prairie marsh wrens, swamp sparrows, rails, and hordes of red-winged blackbirds. Later in the season when the water has receded and the meadow is merely wet, it is a much less busy place.

In a generally forested region, places that are regularly flooded for part of the year but reasonably well drained during most of the growing season commonly bear some type of swamp forest. In the Middle West swamp forests inhabit two kinds of places: flat, poorly drained areas that are underlain by impervious shales or ice-laid clay, or that are really not-quite-extinct postglacial lakes; and wet bottomlands along rivers that periodically overflow their banks and spread out to form wide, shallow lakes. In the general levelness of the Midwestern landscape, many tracts of relatively undisturbed swamp forest still remain; but vast areas of primeval swamp have totally disappeared into farmlands and are detectable now only by their utter flatness and the drainage ditches that cut across them.

Such is the state of the Great Black Swamp that lay in the outlet end of postglacial Lake Maumee, southwest of Lake Erie. This once covered an area bounded roughly in present terms by the cities of Sandusky, Fort Wayne, and Toledo, with arms reaching nearly to Detroit and Cleveland. For years it was an all but impassable obstacle to settlement and even to frontier travel. The only way through it was along the ridges of extinct beaches or glacial moraines. In fact, one can find the locations of many stretches of fossil beach by looking at a modern road map — a road that cuts irregularly across the rectangular survey grid in this part of the country is likely to be an old one that follows what once was the only semi-dry route through the great morass.

On the poorly drained till plains of southwestern Ohio and Indiana, many of the wet upland flats originally bore rather open stands of pin oaks that sometimes reached great size. When such places are drained, there follows a sudden explosion of dense, almost pure stands of young pin oaks. As these grow taller, their lower branches become shaded and gradually die, letting in more light and opening the way for other trees to invade. A special feature of developing pin oak woods is a shrubby ground cover of poison ivy. In the Middle West this noxious and rampant sprawler tends to grow as a rather loosely branching upright shrub quite different from the vinelike habit it assumes farther east.

Other huge swamp forests lay in the valleys of rivers that once carried large volumes of glacial meltwater and are now deeply filled with silt. The Miami and the Scioto in Ohio, the Illinois, the Wabash, and much of the Mississippi, for example, flow through miles-wide valleys that are still subject to miles-wide floods. These, too, are practically all farmland now. The rest was long since cut over for lumber and now bears second-growth woodlands of various kinds.

One piece of virgin bottomland forest that has somehow survived intact is located in Beall Woods State Park in Illinois. This lies along the Wabash, near Mount Carmel and somewhat downriver from the old Indiana town of Vincennes. Here is a tract of almost two hundred acres of true virgin forest. It has been fenced off for the sake of control, but the public is welcomed and there are several nature trails through it.

One of the most impressive things about this forest is the great size of the trees. Nearly three hundred of them have trunks upward of thirty inches in diameter, two of them being more than four feet through. Several of the giants are oaks, including a white oak 117 feet tall. One tuliptree is 129 feet tall, and there are other large specimens of silver and sugar maples, hackberry, and walnut. All together, forty-nine kinds of tree have been found in Beall Woods.

Trees of great size have been found in several other places in the lower Wabash Valley, but nothing remains today that can match the measurements recorded there by Robert Ridgway in the years around 1890. The average height of the level top of the forest then was 130 feet. Many individuals rose to 180 feet, and some sycamores and tuliptrees reached 200 feet, with ninety-odd feet of clear trunk from the ground to the lowest branches. His photographs show such phenomena as black walnuts six feet thick just above the basal swell and sixty feet to the lowest branch, and a group of four tuliptrees with trunks five to seven feet in diameter.

In the southern part of the Middle West the swamp forest is quite diverse and a large number of species are fairly common. This changes toward the north as the trees one by one reach their northern limits. But over all the Middle West and for a long distance beyond, whatever else may be present, the commonest trees of the swamp forest are elm, ash, and soft maples: silver, red, and the compound-leaved maple oddly named box elder. For hundreds of miles the long irregular ribbons of forest that follow the rivers show remarkably little change, even as they traverse a number of quite different kinds of vegetation.

In a swamp forest, a primary fact of life is a wide fluctuation of the water level above and below the soil surface. Through much of the winter and on into spring, water stands at least a few inches deep, much of the time considerably more. This can vary widely from season to season. Sometimes the forest is still flooded after the tree leaves have begun to unfurl. On the other hand, by late summer the top of the soil may be powdery dry and plastered over with a papery mat of last year's fallen leaves. Between times, the small irregularities of the ground surface make a mosaic of little pools and puddles among the slightly higher ridges and hummocks.

A lake plain swamp often develops a considerable amount of peat, since the surface is alternately too saturated and too dry for decay to keep up with the production of dead plant remains. The peat is heavily interlaced with a network of roots, and the ridges formed by large surface roots and the buttressed bases of large trees and any other protrusions become heavily coated with moss.

Such accumulations are possible in a quiet swamp where the water rises and falls gently; but on river bottomlands, the water that rises and falls is likely to be fast-moving floodwater. Consequently another major condition of life on an active floodplain is an unstable, constantly shifting substratum.

River currents are constantly changing, eroding, and laying down silt, now here and now there. Even a stable river keeps taking away from the upstream ends of islands, bars, bends, and meander loops and depositing in the slack water at their downstream ends. Any floating debris that becomes lodged in the river touches off a deposit that may rapidly grow into a new bar or island. As a result, there are always freshly exposed places for plants to colonize, especially as the summer wears on and water in all this region subsides.

On the bare surfaces vegetation develops in a fairly predictable way. The first things to make much of a showing are fast-growing succulent annuals. Seedlings of willow and cottonwood start at about the same time, although they grow more slowly and at first are not very noticeable. Once established, however, the young trees grow at a galloping rate in the fertile alluvium, and in seventy years cottonwoods can be eighty-five feet tall and five or six feet through the trunk.

At the water's edge where the soil is constantly shifting, nothing else can survive the submergence, the burial with silt, and the battering by flood-borne debris that the willows and cottonwoods endure. Nothing matches their ability to heal their wounds and to form new sprouts and send forth roots from damaged or buried stems.

Where the land is more stable and drainage is slightly improved, whether by building up of the floodplain or deepening of stream channels, other trees such as elm, ash, and soft maples can get a foothold. All of these are fairly tolerant of flooding. Silver maples will not merely survive but can actually grow while they are submerged. This has been shown with trees equipped with dendrographs, instruments that measure minute changes in the girth of a trunk. Even in floods that came at the height of the growing season in June, these trees made a small amount of growth every day.

Young seedlings can survive even total submergence during the growing season. Boxes of small silver maples and red maples kept under water for several weeks showed a remarkable indifference to the treatment, for although some shoot tips and parts of leaves were injured, none of the plants was killed, not even after eight weeks. How long it would take to kill them was not determined, for at the end of that time "circumstances dictated an end of the trial."

Once conditions on the floodplain settle down enough to allow a good ash-maple-elm forest to develop, other characteristic river-bottom trees begin to appear. It is in a mature floodplain forest that one finds the giant trees of old records. Further developments depend ultimately on improvement in the underlying soil. So long as this is repeatedly disturbed by floodwater, there is little possibility of much change, and the typical swamp forest persists more or less indefinitely.

In many river valleys the flatland stands at two or more distinct levels. The lowest or "first bottom" is the true floodplain. Inundated by high water practically every year, it can properly be considered a part of the river itself. The higher "second bottom" is a relic of an earlier stage of events and is now flooded only in the extremely high water of superfloods. It is normally well drained and has long been cultivated and built upon — one of the major reasons why the greatest floods do so much damage to the works of man.

Under wild or natural conditions great floods, like other natural disasters, help to maintain diversity in the world of nature, producing many kinds of situations where many kinds of things can live. But widely fluctuating water levels accord poorly with the uses man would make of the valleys, and much has been done to tame the rivers. In this region the chief aim of river control has been to even out the extremes of high and low water, partly for the sake of reducing flood damage, but also to maintain dependable navigation. On the Mississippi there are twenty-six dams and locks between St. Paul, Minnesota, and Alton, Illinois, where the Missouri comes in. The lower reaches of the Illinois, the Wabash, and the Ohio are lined with levees, and their channels have been tinkered with to a certain extent. They, too, are interrupted by many dams and reservoirs, as are many other, lesser streams. The Muskingum River of Ohio is an especially well regulated example of this.

For all the draining and diking and damming that have gone on, there are still places where one can see lush, muddy bottomland forest, sometimes intermingled with marsh. Below the dikes at Vincennes on the Wabash, at

Beardstown or Princeton on the Illinois, at Cairo where the Ohio joins the Mississippi, or between McGregor, Iowa, and Prairie du Chien, Wisconsin, on the Mississippi, the river edges are essentially unmanipulated and left to the wild creatures. Indeed, the upper-central Mississippi has been designated as a great wildlife preserve, which provides homes for millions of resident birds and other animals and a flyway for millions of migrants on their annual journeys to and from the lakes and marshes of the north.

13. Man on the Land: A Thumbnail History

THE MIDDLE WEST as we know it today could reasonably be said to date from the end of the Revolutionary War. Before that time the Ohio country and the central Mississippi Valley had been looked upon primarily as a place where the Indian trade could be carried on, or where the start of an easy route to the western ocean and the Far East might yet be found. The French had explored the region, and for decades they had struggled with the British for mastery of it. By 1763 the British had prevailed. But settlement of this vast, fertile land was discouraged, even officially forbidden, a policy that was received coldly by the colonies whose original grants of land extended far beyond the mountains, some of them all the way to the western sea.

When the war ended and the new nation came into being, in all the region from the Appalachians to the Mississippi and beyond, and from the Ohio River to the Great Lakes, there lived only a few handfuls of people. Widely scattered along the waterways were a number of small communities that had aggregated around missions and military or trading posts, which often enough were the same thing. Along the Lakes there were settlements, largely French, at Detroit, Green Bay, Mackinac, and Sault Ste. Marie, and at Fond du Lac (now Duluth) and Grand Portage on western Lake Superior. The last stood at the mouth of the Pigeon River, where the canoe route to Rainy Lake and the far northwest began.

On the Mississippi there was a fort and village at Prairie du Chien. Around the missions at Cahokia and Kaskaskia were little French settlements. St. Louis, just across the river from these, was a lively and rather civilized small town whose life blood was the Missouri fur trade. Vincennes on the Wabash was an important place from early times, and for many years it remained the largest town in the region. On the Miami River in Ohio at the site of present-day Piqua stood Pickawillany, center of the British trade with the Indians and a considerable town. But despite this scattering of inhabited places, the entire country northwest of the Ohio River remained essentially an undisturbed wilderness.

In the treaty that marked the end of the Revolution the national boundary was drawn along the Great Lakes, thence up the old canoe route from Lake Superior via Rainy Lake to the northwest corner of Lake of the Woods. From there it ran "due west" to the Mississippi, which in fact rises just about due south of there and never reaches so far north. For a number of years there was a large vagueness about the geographical facts of this remote corner of the new country. One consequence of this is the odd, detached bit of Minnesota still known as the Northwest Angle, which nicks sharply into Canada and is accessible from the United States only by water or through Canada. This disjunction resulted from the running of the border from the northwest corner, or angle, of Lake of the Woods, as specified in the peace treaty, to the forty-ninth parallel, which was finally set as the boundary after the purchase of Louisiana in 1803. The short line connecting these markers runs due south through Lake of the Woods. The actual source of the Mississippi was not located until 1832, when an expedition under Henry Rowe Schoolcraft finally found it in the lake he named Itasca.

When the thirteen former colonies set themselves up as a

nation, one of the problems that arose was what to do about the land west of the mountains that had been included in several of their original charters. Massachusetts and Connecticut extended "from sea to sea"; both New York and Virginia claimed territory northwest of the Ohio River; and the Carolinas also had claims running far beyond the mountains. This conflict of interests was finally resolved when all the states agreed to cede their Western claims to the federal government, except for certain limited areas reserved by Connecticut and Virginia to use for paying war compensation.

The lengthy and heated discussions about how to handle the territory lying northwest of the Ohio eventuated in two profoundly important ordinances. The Land Ordinance of 1785 spelled out the manner in which orderly opening up of the federal lands should be accomplished. Before it was sold or opened to settlement the land was to be surveyed and laid out into townships six miles square. Each square mile, 640 acres, was designated a "section," and sections were numbered in a specified pattern within the township. Any tract could be identified accurately, both on the ground and in the land office records, by its section number, the number of the township in the north-south direction, and the range of townships in the east-west direction (for example, Sect. 3, T 2 N, R 1 W).

There were certain exceptions to the standard plan of survey, such as the Western Reserve of Connecticut, where townships were five miles square. Some of the other old grants pretty much followed rules of their own. But the basic mile-square grid, with local roads following section or township lines, laid a physical stamp on the landscape that is still strongly visible nearly two centuries later.

A second ordinance, passed by Congress in 1787, provided for the immediate government of this Northwest Territory and specified that when the population warranted, it was to be divided into states that would join the Union on an equal footing with the other states.

Provision for settlement did not immediately open the floodgates to settlers, however. Before the surveyors could be about their work, the Indians had to be induced to sign formal treaties ceding their lands to the United States. While traders had always been welcomed, landseekers were an entirely different matter and were strenuously resisted. Many were the dead frontiersmen relieved of their scalps in exchange for a mouthful of earth. This merely slowed things, however, and region by region the Indians were driven out of their lands.

The earliest settlements in the new territory were along the Ohio River and were managed by land companies organized as commercial ventures. The very first colony was started in 1788, when a group of New Englanders under the name of the Ohio Company laid out the town of Marietta at the mouth of the Muskingum River. This was followed immediately by a group from New Jersey at Cincinnati and a colony of French émigrés at Gallipolis. In 1796 came Chillicothe, founded by tidewater Virginians and still showing their influence in some of the fine old houses.

Soon some forty-five thousand people were living in what is now Ohio, nearly all of them in the southernmost part of the state. They were coming at such a rate that the land company holdings were soon taken up and settlement moved on into the public domain.

In the first years under the Ordinance of 1785 public land had to be sold at auction, with a minimum price of two dollars an acre, and in plots of at least 640 acres. This required far more cash than a frontier farmer ever saw. Consequently, the buyers were almost entirely speculators, and actual settlers had to buy their smaller holdings from a middleman. This offended the frontier spirit, and much agitation went on in the West for cheaper land and some form of credit. As a result the terms of sale were gradually eased and the size of the minimum salable plot was reduced. Modifications through the decades culminated in

1862 in the Homestead Act, whereby a settler could acquire title to 160 acres at the expense of only two or three small fees and the labor of making certain specified improvements on the land.

The first modification of the land law, in 1800, included the establishment of land offices in the West, near the tracts to be sold. This simplified matters for the hordes of small buyers and also made it possible to keep a degree of order in the processes of sale and transfer of legal title. The first offices were opened in Marietta, Steubenville, Chillicothe, and Cincinnati. Others followed as new areas were surveyed. The rate at which operations went on in these places gave us the phrase "a land-office business," which is so deeply embedded in the national idiom that many have forgotten, and others probably never knew, where it came from.

The first waves of settlers came by the easier of the routes through the mountains. At the south, Boone's old Wilderness Road of 1775 led from Cumberland Gap, at the extreme western end of Virginia, northwest to the early settlements in Kentucky. In time this was extended to the falls of the Ohio at Louisville and on to Vincennes. Wayne's Trace, leading northward from Cincinnati to Greenville, dates from the Indian wars of 1792 and has survived to become U.S. 127. Up these roads and up the various tributaries of the Ohio River into what are now Ohio, Indiana, and Illinois moved the second and third generation frontiersmen and small farmers from the backwoods, with few possessions, scant polish of civilization, and a casual and sometimes romantic approach to the business of living. The stamp of their origin in the southern mountains is still strong in the country just north of the Ohio, most clearly perhaps in Indiana.

Other roads led west from the Middle States. Braddock's Trace, originally a military road of the then incipient French and Indian War, carried traffic from Fort Cumberland, on the Potomac in Maryland, to a point near the forks of the Ohio, now Pittsburgh. A little north of it lay Forbes Road, another rough track through the forest.

From these roads the Ohio River and its branches served as the great highway to the newly opened lands. But with floods, ice, low water, and other hazards, the river was often unnavigable. To provide a means of avoiding the problems of river travel, Congress in 1796 voted an appropriation and authorized Ebenezer Zane to cut a new road from Wheeling in Virginia across what is now Ohio to Limestone (now Maysville), Kentucky. Zane's Trace was merely a narrow gash through the tall forest; but where it crossed the larger rivers, important towns sprang up — Zanesville, Lancaster, and Chillicothe.

At the north, by the turn of the century the Great Genesee Road extended across New York State from Albany past the Finger Lakes to Buffalo. From there a road of sorts ran west along an Indian trail that followed an old postglacial beach ridge. From the Pennsylvania line to Cleveland a "girdled road" — made by removing a ring of bark around each tree trunk — offered a track through the brush among dead but still standing trees.

Over all these primitive roads, rutted, muddy, and strewn with stumps, frontier-bound settlers poured in a heavy stream. And in the time-honored way, the public set up a loud clamor, demanding that the government should both construct and thereafter maintain an improved road to the West.

Toward this end Congress took the first steps in 1806, and by 1817 the Cumberland or National Road was open to travel, following Zane's old route from Cumberland, Maryland, to Wheeling. This was a substantial road, broad and smoothly graded, with a high crown in the center for drainage and a thick surfacing of crushed stone. It promptly became a major artery, and in time it was extended across the whole Northwest from Wheeling through Zanesville to Columbus, Indianapolis, Vandalia, and on to the Mississippi opposite St. Louis. It has remained an important route of travel, metamorphosing through U.S. Highway 40 into Interstate 70 in the current generation of highways. Its countryside reflects the Middle Atlantic origin of the first settlers, and many old houses along the National Road could as well date from early days in eastern Pennsylvania as from the pioneer times of Ohio.

In the North, settlement began more slowly. After the Revolution the British, intent on keeping their hold on the Indian trade, refused to evacuate their posts around the Great Lakes and on the western rivers and did nothing to restrain their Indian allies in resisting settlement. Not until after the War of 1812 were these hazards removed.

First in the line of northern settlement was the Connecticut Western Reserve, lying between Lake Erie and the forty-first parallel and extending 120 miles west from the Pennsylvania boundary. The state of Connecticut retained a 25-mile strip at the western end of the Reserve to use, in the absence of cash, for compensating those of its citizens whose property had been burned by the British during the Revolution. From this use it became known as the Firelands. The rest of the Reserve was sold in 1795 to the Connecticut Land Company. The following year the company sent out a party under Moses Cleaveland to "quiet" the Indian claims to the land and survey it as far as the Cuyahoga River, at whose mouth, exactly in the middle of the shoreline of the Reserve, they were to lay out a townsite. For a while this town was only a geographical concept, for in 1800 it had a total population of seven. The land west of the Cuyahoga was not ceded by the Indians until 1805; and in that year the Firelands tract was at last distributed among the claimants to it.

Settlement in the Reserve was very slow at first. But many families in the East suffered heavy losses in the War of 1812, and after the bitter winter of 1816 — "eighteen hundred and froze-to-death" in New England — a large migration from Connecticut into the Reserve began. The next few decades saw the growth of a veritable New Connecticut in all senses of the word. Some of the towns founded in those years are still very New Englandish in appearance, with handsome white houses and church set around a village green. A striking example is Hudson, which long retained its Yankee ideals and ways. Its Western Reserve University served for many years as "Yale in the West." Talmadge, Painesville, and, in the Firelands, Norwalk and New London still preserve the New England look. Elsewhere, rapid urban and industrial developments have totally obliterated all evidence — physical, social, and intellectual — of the origins in New Connecticut.

In contrast to the fairly civilized life in even the earliest stages of towns, life for the pioneer farmer in this region was hard, raw, and lonesome. Whereas in the southern part of Ohio the early farmers had arrived and settled in neighborly groups, in the Connecticut Reserve they usually came as single families. This was the result of the way the Connecticut Company disposed of its land. Its entire holding was sold on shares before any of it was allotted to specific purchasers. While one could buy any size tract he wished, the location of individual holdings was determined in a grand lottery. As a result, the first farms were scattered at random all over the area, most of them far from neighbors and further isolated by the widespread swampiness of the land and the impassable condition of what roads there were. Travelers' accounts of the pale, scrawny, ague-ridden pioneer, struggling to eke a living from the dense forest, his children growing up unschooled and often illiterate, do not make a pretty picture; and there is little reason to believe that life at that time and place offered any advantages beyond independence, owning one's own land, and a hope that the future would be better.

The success of pioneer farming was heavily based on the labor of man, woman, and beast, usually oxen. If his land was reasonably fertile, a man could start with little but a few cattle and a few elementary tools — an ax, a plow, and a gun — and within his lifetime create an inheritance for his growing family. Basic to this possibility were the abundance and usefulness of wood, the abundance of wild game in the forest, and the ability of the corn plant to grow and yield something of a crop on land that had received only the most primitive cultivation. From the trees came logs for the first cabin, rails for fencing, and dozens of ingeniously made articles of furniture, household implements, and farm tools. For the first year or two wild game often provided the only food that the family had for months on end. Such cattle and hogs as they had ran wild to forage on their own in the woods. The ability of these animals to look after themselves was much more important than their quality or productiveness.

The first corn crop was usually planted in a "deadening" in the woods. Here the underbrush was cut, but the larger trees were left standing, being merely killed by girdling. A rough plowing broke up the tough mat of roots, and a few corn grains were planted in holes cut into the debris. Pumpkins and gourds, those plants of a thousand uses, were interplanted and allowed to run among the cornstalks. Beyond that the crop pretty much took care of itself while the farmer busied himself with other things. While the grain was sprouting and again at the ripening of the crop, women and children often did guard duty to chase off the crows, squirrels, and raccoons that always know a good thing when they see it.

Year by year the trees that had at first been left standing had to be cut and their stumps pulled and their larger roots grubbed out. All this trash had to be disposed of in some way, and this was usually done by burning. The resulting ashes served as fertilizer for the corn crop. Gradually the raw humus of the forest and the woody remains of smaller roots decayed to make a mellow soil that was further improved by the regular addition of manure. It took about seven to ten years for the raw forest soil to develop to a condition that would yield a good crop of wheat, which was usually the first cash crop.

For a long time farming was essentially self-sufficient and independent. The few large landowners who came west expecting to operate large estates with the help of hired labor found that since everyone was busy on his own land, there was nobody available to be hired. The perpetual scarcity of labor, with each man doing his own work and master of his own life, made a large contribution to the egalitarian spirit that pervades our frontier history.

The fertile soil of this region yielded an abundant produce when it had been brought under proper cultivation, and before long there were surpluses beyond the needs of even the burgeoning frontier towns. But where was a farmer to market his surplus crops? Shipping back to the East was far too expensive, for the cost of freighting over the wagon roads was greater than the final market value of the goods. Produce that could walk to market on its own feet might sell at a profit, however, and thousands of cattle and hogs made the slow journey over the mountains under the care of drovers. The many large herds on the roads were one of the impediments that the westward-moving settlers had to contend with. There was some profit to be had by concentrating various grain crops into the form of whiskey before they were shipped. This was such a widespread practice that when in 1794 Congress imposed an excise tax, the farmers of western Pennsylvania raised a violent protest that became known as the Whiskey Rebellion, and the tax was repealed in 1802.

For decades the chief outlet for farm produce was down the Ohio and the Mississippi to New Orleans, whence it could readily be shipped to the East, or for that matter to Europe. Many a farmer took his own crops to market this way, floating them down on a flatboat that was often less a boat than a primitive raftlike affair with no control but a steering oar. With no motive power but the river current, his journey was slow but cheap. Selling his crop at a decent profit and getting back home with the proceeds was another matter, and all too few of those who tried it did very well by the deal. Apart from the adventure of the trip, it was better to sell to a local merchant, who might be the keeper of the nearest country store. Such commerce between farmer, local merchant, and larger distributor was an important factor in the growth and prosperity of many river towns.

As commerce on the rivers developed, a variety of craft came into use. Among these were keelboats, long narrow boats that resembled overgrown rowboats with a deck added. These floated downstream on the current but could also, at considerable expense of labor, be poled, towed, or cordelled back upriver.

The first steamboat appeared on the Ohio in 1811, and thereafter its numbers increased rapidly. Year by year they grew larger, faster, and grander; but always they were flat-bottomed vessels with the shallow draft that was necessary for the many shallow places on their routes of travel, especially in the low-water season. Navigation was complicated not only by the constantly changing currents, channels, and shorelines, but also by the great differences between low water and high, a difference of fifty feet within the average year on the Ohio. Highest and fastest water ordinarily came with the spring floods between April and June, and the melting snow of late winter brought another somewhat lesser flood stage. In late summer between August and October the river was vastly reduced in all its aspects. The "falls" of the Ohio at Louisville, originally a two-mile cascade over rocky ledges, could be run at high water; but they were sufficiently hazardous for a canal finally to be built around them on the Kentucky side in 1830. Not until 1879 did Congress take action to build the first dams and locks on the Ohio River. Since then the Ohio and most other major rivers have been thoroughly tamed by engineering works.

On Lake Erie the first steamboat, the "Walk-in-the-Water," appeared in 1818. This inaugurated a flood of westward migration via Buffalo into the more northern parts of the Old Northwest. Sightseeing trips by boat became popular, and by the 1840's many easterners and even European visitors were making a fashionable tour of the West that took them down the Ohio, up the Mississippi, and home by way of the Great Lakes.

Except for the National Road and a few other turnpikes that were surfaced according to the method of John Loudon McAdam, most roads were in a perpetually atrocious condition. Many of them served more to mark the route than to ease the conditions of travel. The more heavily used a road was, the deeper were its ruts. Lesser tracks along the township lines no doubt deserved their description of "stump-cluttered mud bogs." Lumber being one of the cheapest commodities on the frontier, there was a period around 1850 when plank roads were built with great hope and enthusiasm. At one time there were six of them radiating from Chicago. The planks rotted rapidly, however, especially in the wet places where they were most needed, and in the absence of constant maintenance, the usual situation, they soon disappeared.

In northern Ohio the ridge road along the lake shore from Buffalo through Cleveland to Sandusky ended for many years at the edge of the Black Swamp. This large, forested morass was covered most of the year with several inches of water and effectively isolated Michigan from nearby Ohio, delaying the development of that area for a long time. When a military road was finally put through to Detroit in 1827, it took four years to build the thirty miles across the swamp. After the swamp was drained in the 1850's it became a most fertile expanse of farmland. The level terrain and the ubiquitous drainage ditches still bear witness to the original condition of the area.

The year 1825 marked the beginning of a new phase of western development. The impetus for this was the opening of the Erie Canal across New York State from Albany to Buffalo. Even the small, cramped canal boats moving at the four-mile-an-hour pace of a tow horse or mule made possible a vastly increased traffic of immigrants moving westward and agricultural produce, largely wheat, moving east. Shortly after that the Ohio Canal was built connecting Lake Erie with the Ohio River via the Muskingum and the Scioto. This gave the first important boost to the village of Cleveland, which lay at its northern terminus.

Many canals were acts of hope, for in 1815, when the Ohio legislature authorized both the Ohio Canal and the Miami Canal farther west, about five-sixths of the state was still a primeval wilderness. Other canals were eventually dug at all the major portages between the Great Lakes and the Mississippi River system. Although the canal era lasted little more than two decades, it gave the Middle West its first really adequate transportation and brought rising prices and a measure of prosperity to the western farmers.

Many of the old canals have disappeared and the routes of others are now occupied by roads or railroads. Only two are still functioning, both of them in much moderized form. The New York State Barge Canal supplants the Erie Canal, and the Chicago Sanitary and Ship Canal connects Lake Michigan with the Illinois River. Surviving bits of the old Whitewater Canal at Metamora, Indiana, and the Ohio Canal at Canal Fulton have been restored to their original condition, complete with locks and gates, as museum pieces.

Even while the canal-building fever was at its height, railroads began to make an appearance. The first one west of the Alleghenies was built in 1836 and extended thirty-three miles from Toledo to Adrian, Michigan. In the early years, railroads were regarded chiefly as a means of connecting navigable waters. In this region most of them ran north and south, and it was not until 1852 that the first train directly from the East arrived in Chicago.

Meanwhile, as settlers poured into the Northwest Territory and parts of it became reasonably well inhabited, new states were organized and admitted to the Union: Ohio in 1803, Indiana in 1816, Illinois in 1818. In those years the population of all three was heavily concentrated in the south, and their northern parts were still empty wilderness.

When Michigan Territory was organized in 1805, most of its sparse population were descendants of the early French, and the 750 people in Detroit still lived in the same style as the habitants along the St. Lawrence in Canada.

By the later 1820's, however, settlers were flooding into the northern parts of the region as fast as roads, canals, and steamboats could carry them. A road through the Black Swamp finally connected Detroit with the settled parts of Ohio; and the Territorial road built across southern Michigan in 1833, although it was intended primarily to provide communication between Detroit and the frontier military posts farther west, served as a great immigrant route and encouraged settlers to take up land in the fine country it traversed.

The sizes of towns in 1830 show the relative state of development of the different regions. At that time Cincinnati was by a good deal the biggest town in the West, with almost twenty thousand people. Cleveland was just beginning to benefit from the Ohio Canal and had grown to a thousand. Detroit, newly emerged from its long history as a French frontier post, was approaching two thousand. Chicago was a village of less than a hundred and fifty souls clustered around Fort Dearborn on the empty prairie.

By this time the advancing frontier had encountered the first of the prairies in Ohio and southern Michigan. The oak openings along the prairie border offered a sort of transition from life in the forest. But the wide open, completely treeless prairie beyond presented entirely new problems. The whole logistics of frontier farming had evolved on the basis of an abundance of wood, and this the prairie farmer had to learn to do without. Another problem was water, for there is much less surface water on the grassland than in the better watered forests; consequently, digging a well was often a first and urgent piece of business.

Breaking new land on the prairie presented still other unfamiliar problems. In the first plowing, the tough sod had to be undercut. For this purpose a new type of sod-breaking plow was evolved that cut a shallow but very wide furrow. "Sod-busting" had to be done early in the season while the soil was still moist enough to be workable, but late enough so that it would soon dry out and curb the resprouting of the vigorous prairie grasses. It took several yoke of oxen to pull a plow through the virgin prairie sod. Things improved with the appearance of John Deere's steel plow, which he first devised from a millsaw blade in 1837. The heavy prairie soil did not stick so tenaciously to the smoothly polished surface of the steel share as it did to the old iron-covered wooden ones that had served the earlier frontier farmer.

Fortunately, on the prairie as in the forest, freshly broken land yielded an adequate subsistence crop of corn in its first year. All that was necessary was to cut a few slits in the overturned sod with a hatchet and drop in the corn grains. It took only a year or two for the remains of the sod to rot, and very soon a cash crop of wheat could be had.

Wheat as a principal crop seems to have followed the frontier west all the way from the beginnings in New England. Each region that was opened seemed better adapted to wheat than the last one, and time and again competition from newer lands diverted the more settled country to some other crop that proved to be more profitable. Thus, much of the Middle West changed from pioneer corn patch to well-developed wheat country before it achieved its present status as Corn Belt.

Settlement of the prairie region of northern Illinois and nearby Wisconsin was delayed until after the end of Black Hawk's War in 1832. This war was one more case of Indian resistance to ceding their lands, and it ended as usual with the ousting of the native inhabitants. Development of this area took place during a period of furious land speculation, chiefly involving absentee capital. Nothing like it was seen again until the Florida boom of the 1920's. There was wild buying and selling; many a fortune was made and lost, at least on paper; and much finagling took place about where townsites and especially county seats should be located. Riding high on all this was the boom town of Chicago, whose vital statistics are spectacular and irresistible:

1832 — population 150
1833 — population 2000
1837 — incorporated as a city of 4117
1850 — population 30,000
1870 — population 300,000; the greatest railroad center in the world
1890 — population 1 million

While Chicago was burgeoning from a lonely hamlet into a metropolis, the rest of the Middle West was becoming inhabited country. The story of man on the land becomes very complex, and many things happened at once that make their mark on the landscape of today.

Farming has continued to be a conspicuous part of the scene in the areas that were originally prairie or deciduous forest. Most attempts at cultivating the soils that bore evergreen forest have resulted in failure. One exception is the use of bog land for growing cranberries. Practically all the cranberries raised in this country that do not come from Plymouth County, Massachusetts, come from the bogs of Wisconsin and Minnesota. The sources can be identified by the trade names adopted by the regional growers: Ocean Spray identifies the one, Eatmor the other.

The true prairie is indeed the land where the tall corn grows. In fact, between midsummer and harvest time a traveler's view of the landscape is sharply limited by seemingly endless rows of cornstalks that reach considerably higher than the top of his car. Visibility is better in places where the huge fields, a mile square and more, are in some non-corn stage of the crop rotation sequence.

Where valley floors or old lake bottoms offer especially fertile soil, produce is grown for canneries or freezeries, or truck crops for the urban masses. Other regions of specialized agriculture lie to the east of Lake Michigan and south of Lake Erie. Downwind of these large bodies of water the weather is moderated, so late spring frosts are less common and extreme winter temperatures are less severe than inland, and these areas have developed into major fruit-growing regions. Grapes, peaches, and cherries are some of the most conspicuous crops.

The nature of farmsteads varies from place to place. Farmhouses may be of any vintage from the early 1800's to this year. The age of the oldest houses depends on where you are, for it wasn't until 1839, for example, that the first road was laid out in Iowa. That was when Lyman Dillon plowed a hundred-mile-long furrow across the prairie to guide the road builders from Dubuque via the site of Iowa City to the Missouri line. North and west of that, century-old buildings are rare indeed. Farmhouses of modern construction indicate that this is a region of current prosperity, although many houses of all ages bear little resemblance to the nostalgic image of a farmhouse that stems from New England.

A good deal of the corn grown in the Middle West is converted to meat. Big families of small piglets grow rapidly and efficiently into large hogs.

There seems to be a correlation between the plantings around a farmhouse and the type of original vegetation on the site. In the deciduous forest region there are often huge old sugar maples and other large trees still in a flourishing condition. In the Prairie Peninsula and the prairie border region, the original old trees that may survive from a prairie grove or oak opening, usually supplemented with others such as the ubiquitous blue spruce, often show the ravages of time, with their tops blown off or large boughs missing or other signs of decrepitude. In fact, the state of health of large farmyard trees might be a good clue to the original vegetation of the place, if one were to look into the matter in a systematic way.

On what was originally open prairie, farm buildings are often completely buried in a plantation of trees, or at least shielded on the north and west sides with a dense windbreak planting. Driving along a road thereabouts, looking to the south or east you may see only mounds of trees, many with windmills protruding above them. To see the buildings, you must look toward the north or west. Going from the sun and wind of the open fields into the calm of a tree-shaded farmyard gives a strong sense of welcoming shelter in the prairie land.

In such level farm country, a dozen or more farmsteads are often visible at one time around the horizon. Even though they are scattered some distance apart, there is not the sense of isolation that one feels at a lone farmhouse in hilly, wooded country, which may in fact be more densely populated. Farther west, the farms are bigger and the houses farther apart. But throughout the Middle West as anywhere else, as the size and quantity of machinery continues to increase, farms grow ever larger, yields become higher, and farmers become fewer and fewer.

While agriculture is based heavily on sediments deposited by the ice sheets, the underlying bedrock has yielded mineral products of many kinds. Indiana limestone and Ohio sandstone are widely used building materials, and a goodly proportion of sidewalks, at least in northern Ohio, are made of large, smooth sandstone slabs (a marvelous surface for roller skating). Where the Niagaran formation crops out through the surface, as for instance west of Sandusky, Ohio, or at Alpena, Michigan, there are quarries and cement and gypsum plants. The granite of the Canadian Shield is quarried where it is exposed in the trench of the Minnesota River, and there are coal mines in southern Illinois and eastern Ohio.

One of the few picturesquely old regions of the Middle West is the lead-mining country of the Driftless Area in northwestern Illinois and nearby Wisconsin and Iowa. The towns of Galena, Mineral Point, and Dubuque all grew rapidly during the "lead rush"; and in 1836 half the entire population of Wisconsin lived in the lead country. Many of the people were southerners who came up the Mississippi by boat. Others were mine workers imported from Cornwall. Some of the miners' old-world type of stone cottages remain as mementoes, and Cornish pasties appear on local restaurant menus on certain days of the week. The larger white-painted frame houses of more prosperous inhabitants still stand along old-fashioned tree-shaded streets. These are towns that have been spared from change by the fact that since the lead ores gave out in the late 1840's nothing else has happened to submerge them under the wave of later urban expansion that has obliterated so much of the old in this part of the world.

It was known from earliest times that there was copper somewhere in the Lake Superior Country, for the Indians used it in its "native," unsmelted metallic form. One of the various duties assigned to the Cass expedition on its journey to Lake Superior in 1820 was to investigate the source of this metal. They did find a vein of it thirty miles up the Ontonagon River from the lake, but the amount of metal proved to be much smaller than rumor had suggested. The more important deposits of copper in the form of oxide ores on the Keweenaw Peninsula nearby were not found until a number of years later. It was in 1844 that mining of this ore began at the town of Copper Harbor. For forty years the great bulk of the copper mined in this country came from the Keweenaw mines. Not until 1887 were they surpassed by the deposits at Butte, Montana.

The last veins of high-grade copper ore gave out in the 1920's. The effect of this on the local economy was so disastrous that in 1930 about eighty-five percent of the entire population in the Keweenaw region was on relief. Some of the old towns with abandoned and dilapidated buildings still stand as a most depressing sight. Recent years have brought new demands for copper and new methods of extraction, however, and some of the towns show a new and hopeful prosperity.

Much more extensive than lead or copper mining ever was is the mining of iron ore that still goes on on a large scale. The initial discovery of Lake Superior iron was made in the Marquette-Ishpeming region of Upper Peninsula Michigan, not very far from the Keweenaw. That was in 1844. The discoverers were a group of surveyors running one of the last government land surveys in the old Northwest Territory. The huge ore bodies of Minnesota in the Vermilion and Mesabi districts northwest of Duluth were discovered later, in 1884 and 1891.

The vast, intertwined industrial complex of iron ore, coal, shipping, and steel that bulks large in the present life of the Lakes States arose from a number of small separate strands. There was an ironworks in Cleveland as early as 1840; its source of ore was bog iron from deposits scattered around the Western Reserve. The first Lake Superior ore arrived at this establishment in 1852, in six barrels carried aboard the vessel *Baltimore*. Coal was available in the town at this time, but it was sold by the bushel and used only for domestic heating. Iron was still being smelted with charcoal, most of it close to the mines. The effect of this procedure on the forests was appalling, for each furnace each year consumed the wood from a minimum of fifteen hundred acres.

In the course of the 1850's several things happened that

interacted to yield large consequences. For one thing, coal replaced charcoal in the iron furnaces, shifting the source of carbon from the primeval forest at the north to the mines southeast of the Lakes; for another, the digging of a canal around the Sault rapids made it possible to ship the ore directly from anywhere on Lake Superior to ports on the lower Lakes. This coincided with the building of railroads which could carry large loads of ore from the iron mines to the northern harborsides, as well as large loads of coal from the coal mines to the lower lake ports. This concatenation of circumstances was responsible for the location of steel mills along the shores of the lower Lakes.

The story of the industrial empires that were constructed in the latter part of the nineteenth century is a fascinating one, but not really part of a book about the landscape. The physical manifestations of those empires, however, are very conspicuous in the landscape of today.

The first iron mines were all open-pit operations, until in the 1880's the Michigan mines began to go underground. Those in Minnesota still consist of vast open basins dug into the earth, so large that only the minute appearance of huge trucks and earth-moving machinery working down within them makes their actual size comprehensible. There are viewing platforms set up with explanatory signs at several places in the region where visitors can safely observe and can learn something of the processes that go on.

At the other end of the system are the steel mills, concentrated along the south shores of Lake Michigan and Lake Erie. Here a smoky pall by day and a bright glow in the night sky are familiar parts of the scene when the economy is in full swing.

In between is the great link of shipping on the Lakes. Along the docks where the trains come in there are giant conveyor systems that can tip a whole railroad car on end and empty it into a ship's hold. Unloading the bulk cargoes is also totally and efficiently mechanized, and an entire boat can be emptied in a matter of hours.

Anywhere on the Lakes, from Buffalo to Chicago to Du-

The bulk-carrying vessels indigenous to the Great Lakes have an unmistakable form. This ore boat is "going up light" in the St. Mary's River, approaching Sault Ste. Marie, Michigan.

luth, from the going out of the winter's ice until navigation closes in late autumn, the long, slim ore boats pass back and forth. For some reason the craft indigenous to the Great Lakes have always been "boats," however large they may be. Not until the St. Lawrence Seaway opened to oceangoing vessels were there "ships" on the Lakes.

The form of the cargo boats that ply the Lakes has evolved to meet the special circumstances of their habitat and use. They are narrow to fit the confines of river channels and the locks at the Sault. The latter also limit their length, although this has periodically been changed by the opening of new and larger locks. The middle segment of the boat is entirely occupied by bulk cargo space and covered with great flat hatches. The machinery and living quarters are concentrated in relatively small segments fore and aft. High at the ends and long and low in the middle, with a smokestack located close to the stern, these vessels could be recognized instantly anywhere in the world.

Widespread over the landscape of the upper Lakes is evidence that lumbering has been a major industry. The first

Present-day logging is highly mechanized. Houghton, Michigan.

steam sawmill in Michigan was set up on the Saginaw River in 1834. Large-scale lumbering was soon under way, although for a few more decades New York State would still be the nation's chief source of timber. When the eastern forests were largely depleted, the wave of reckless cutting swept on into the Middle West, its peak coming in the 1870's in Michigan, the 1880's in Wisconsin, and the 1890's in Minnesota. It was the great pines of the northern forest that were most valuable for lumber, yielding large boards of clear, smooth-grained wood.

These were the years when the prairies were being settled, and not only farm buildings, but all the houses, shops, and other places of commerce and industry in that part of the country were built of wood. The makings of all these structures came down the rivers from the north, mostly in the form of large rafts of logs. Many a river town rose to prosperity on its sawmills and the distribution of cut lumber. A good part of southeastern Iowa, for example, was built of lumber processed at Keokuk and shipped on to the hinterland.

The history of the lumbering years is strewn with hair-raising incidents of runaway forest fires. These had always been a part of the life of the forest, as we have seen in earlier chapters. But now all through the forest there were concentrations of people living in towns where even the sidewalks were made of wood, and the human disasters that accompanied the forest holocausts were appalling.

One of the worst of the many bad fires broke out the same day as the great Chicago fire, October 8, 1871, and swept for forty miles along Green Bay in Wisconsin. It completely wiped out the towns of Peshtigo, Holland, and Manistee and devastated six whole counties before it came to an end. Historical markers along the roadsides in the area spell out some of the incidents of horror. In the cemetery at Peshtigo there are many graves of individual people who died in that disaster. There is also a mass grave of 350 unidentified men, women, and children. There were other fires almost as bad, if less famous, all the way from Saginaw to the edge of the prairie in Minnesota.

By the turn of the century the huge stands of virgin pine were essentially gone, although the last of the really big trees were not cut until the mid-1930's. That was in Michigan, around Whitefish Bay near the Sault. The problem then arose of what to do with all the cutover, burned-over land. The answer that presented itself was to sell it to settlers for farms, especially to farmers coming as immigrants from Europe. It had always been assumed that "the plow follows the ax," and that when the lumbering ended, as in the older country farther east agriculture would naturally take its place. This was one reason for the lack of concern about fire on the cutover land. After all, the hard labor of removing the trees would not be necessary, and for that the settler should be grateful. But the sight of the ravaged land must have caused a heart-wrenching jolt in the bosoms of many a newly arrived, travel-weary Scandinavian family.

Subsequent developments were not much better than first

impressions. The pine land was ill adapted for agriculture under any circumstances, and the climate did nothing to mitigate the difficulties. Most attempts at farming the cutover lands were one-generation affairs at best; and the farm depression of the 1920's, even before the more widespread Great Depression of the 1930's, saw hundreds of hopeless enterprises given up. Abandoned farms added their desolation to that of stump lands and abandoned sawmills. Since there were no buyers at any price, the counties and states came into possession of a good deal of then worthless real estate by reason of tax delinquency.

Through all these years of exploiting the region's natural resources, intertwined with every other development was that of transportation. For a long time roads, rivers, and lakes largely determined where men would go and what they could profitably do. In their infancy, even the railroads were regarded merely as connecting links between old, established routes of travel. But in the 1850's Congress began the practice of making large grants of land from the public domain to the railroads for the purpose of promoting new lines that would open the country to settlement. It was then that truly large construction projects became economically possible, and the rails began to push far out into still unoccupied country. On the prairie, the tracks could often be laid directly on the undisturbed ground. This made for cheap and rapid construction, and it also saved a number of pieces of virgin prairie for the edification of future naturalists.

By the 1860's the population had begun to follow the railroads, rather than the other way around; and the railroads were doing all they could to promote this development. Along their routes they laid out towns at what seemed appropriate intervals. The land awarded by Congress they sold either to middleman speculators or directly to buyers who would occupy the towns and to settlers who would farm the land and generate business for the inhabitants of the towns to transact.

One more thing was needed in a hurry for this system to succeed, and that was large numbers of settlers, even more of them than the rest of the growing nation could provide. To meet this need the railroads sent recruiters to the source of supply, the docks of port cities, where hordes of foreign immigrants were arriving daily in those years, and even to their European homelands. Sometimes the recruiting agents accompanied their settlers from the disembarking place all the way to the land they finally chose to settle on. This benefited both parties: it helped the newcomers, who knew neither language nor customs of their new country, and it prevented distraction and loss to the beguilements of competing recruiters.

For not only the railroads, but also the governments of the newer states — Michigan, Wisconsin, Minnesota, and the Dakotas — were aggressively seeking settlers. The countries of northern Europe were flooded with broadsides, pamphlets, and books setting forth the advantages of the various regions. Each locality was described in vigorous, glowing prose as *the* promised land, the very Eden of all America.

To the farmlands of the Middle West the great bulk of settlers came from Germany and Scandinavia, the Germans somewhat earlier, beginning in the 1830's. The biggest influx from Sweden and Norway came later, when the more northern areas were opening up. Many Finns came early to work in the copper mines, and more of them later to man the northern lumber camps. A group of Icelandic fishermen, the only such in this country, settled on Washington Island at the end of the Door Peninsula in Wisconsin in 1870. The Dutch who settled around Holland, Michigan, came in the 1840's. And so it went. All this still shows in many small ways a century later, even after generations of the traditional American mobility. An interesting pastime for travelers is to read the names on roadside mailboxes and guess at the history of the region they are passing through.

The major influx of immigrants from central and southern Europe, the Poles, Czechs, and various Balkan peoples, settled in the burgeoning cities of the late nineteenth century. Still later came the Italians. And World War I saw the first mass migration of black people from the rural South to the cities of the North.

These, of course, are only the highlights, and there are people of every conceivable national origin in the Middle West today. Many studies of ethnic origins have been made for a variety of purposes, and one of the striking findings is that a native Midwestern adult without at least one foreign-born grandparent is still somewhat of a rarity.

While the countryside was filling up, villages were growing into towns and towns into cities, some at an explosive rate nearly comparable with that of Chicago. By 1890 the population of Ohio was already predominantly urban. With this rapid growth and the overriding geographical expansion that went with it, little of the picturesquely old remains in the cities of the region.

As of now the Midwestern landscape is either intensively farmed, highly industrialized, or densely suburban. Only the north-woods country is still in the process of turning into something specialized.

The freshwater empire of rivers and Great Lakes continues to thrive. The Ohio and the Mississippi are important arteries of barge-borne traffic, as is the canal connecting those great river systems with the Lakes at Chicago.

Another important link in the waterways is the canal with its several locks that bypasses the rapids at Sault Ste. Marie. This is indeed a heavily traveled thoroughfare, for it carries a large traffic of iron ore as well as bulk cargoes of grain from the northern prairies and plains. Since the opening of the St. Lawrence Seaway and the enlargement of the Welland Canal, which circumvents the falls at Niagara, the Sault traffic is widely international; and along with the grain and ore boats there are general cargo ships from such faraway places as Norway and Greece tied up at the docks in Duluth, in the heart of the continent.

At the narrow connecting passages between the Lakes —

the Detroit River and the St. Clair, to some extent the Straits of Mackinac, and especially in the St. Mary's River and at the "Soo" locks — one can sit for hours watching the procession of ships and "boats." At the Sault an observer stands almost within arm's reach of people aboard the passing vessels, particularly while they are in the process of locking through. Here again there are specially built observation platforms for visitors, and it is worth a few hours' stop to watch the passing spectacle.

One rapidly expanding use of the remaining unfarmed, unurbanized parts of the region is recreation in all its aspects. For hunters and fishermen, boaters and skiers, canoers and natural history enthusiasts, the myraid lakes and streams and the forests of the north provide an escape from the conurbations where most people live today.

Resort areas vary tremendously in their atmosphere, as they do anywhere. A few of those in the Middle West have a long history. As early as 1761 Mackinac was a favorite visiting place for the friends and relatives of British officers who were stationed there; and before the Civil War many southern planters came up the Mississippi by boat to the St. Paul region for the summer. A few places retain the prosperous but low-keyed look of city taste and city money applied continuously for many generations. Among those known to the writer are Lake Geneva and Bailey's Harbor, both in Wisconsin, and much of the eastern shore of Lake Michigan from about Muskegon north. The summer resorts of the Middle West would make an interesting subject for a study of social history.

Like any other well-populated area today, the Middle West has its massive problems, such as pollution of air and water, urban congestion and ugliness, and suburban sprawl, as well as pockets of rural poverty. In spite of all this, and in spite of its levelness and its generally unspectacular landscape, it has much of interest for the informed traveler's eye, and a large part of it still remains a rich and beautiful country.

Notes
Index

Notes

Chapter 3

Page 37. The quotation is from N. M. Fenneman, *Physiography of Eastern United States,* McGraw-Hill, New York, 1938, p. 515.

Page 44. Quoted from J. L. Hough, *The Geology of the Great Lakes,* University of Illinois Press, Urbana, 1958, pp. 249–250.

Chapter 4

Page 59. W. S. Cooper and H. Foot, "Reconstruction of a Late Pleistocene Biotic Community in Minneapolis, Minnesota," *Ecology,* Vol. 13 (1932), pp. 63–73.

Page 60. This thick, buried soil is described by J. Voss, "Forests of the Yarmouth and Sangamon Interglacial Periods in Illinois," *Ecology,* Vol. 20 (1939), pp. 517–528.

Page 64. L. S. Dillon discusses the muskoxen in "Wisconsin Climate and Life Zones in North America," *Science,* Vol. 123 (1956), pp. 167–176.

Page 69. R. M. Mengel, "The Probable History of Species Formation in Some Northern Wood Warblers (Parulidae)," *Living Bird,* Vol. 3 (1964), pp. 9–43.

Page 70. For details of this fascinating proposal, see Martin's paper, "The Discovery of America," *Science,* Vol. 179 (1973), pp. 969–974; also James E. Mosimann and Paul S. Martin, "Simulating Overkill by Paleoindians," *American Scientist,* Vol. 63 (1975), pp. 304–313.

Page 71. Such job opportunities are described by Paul S. Martin in "Pleistocene Niches for Alien Animals," *BioScience,* Vol. 20 (1970), pp. 218–221.

Page 73. This paper is by E. A. Bourdo: "A Review of the General Land Office Survey and of Its Use in Quantitative Studies of Former Forests," *Ecology,* Vol. 37 (1956), pp. 754–768.

Chapter 5

Page 80. R. W. Chaney, "A Comparative Study of the Bridge Creek Flora and the Modern Redwood Forest," Carnegie Inst. of Washington, Publ. 349 (1925), pp. 1–22.

Page 85. This expedition is described by K. Chu and W. S. Cooper in "An Ecological Reconnaissance in the Native Home of *Metasequoia glyptostroboides," Ecology,* Vol. 31 (1950), pp. 260–278.

Chapter 6

Page 105. The study in the Nicolet National Forest was done by D. K. Maissurow and reported in "The Role of Fire in the Perpetuation of Virgin Forests in Northern Wisconsin," *Journal of Forestry,* Vol. 39 (1941), pp. 201–207.

Chapter 7

Page 114. The changes around 1400 are described by F. E. Egler in *Man's Role in Changing the Face of the Earth,* W. L. Thomas, ed., University of Chicago Press, Chicago, 1956.

Page 115. For the history of the Itasca Forest, see S. H. Spurr, "The Forests of Itasca in the Nineteenth Century as Related to Fire," *Ecology,* Vol. 35 (1954), pp. 21–25. Also S. H. Spurr and J. H. Allison, "The Growth of Pure Red Pine in Minnesota," *Journal of Forestry,* Vol. 54 (1956), pp. 446–451.

Page 119. For the story of Kirtland's warbler, see H. W. Mayfield, *The Kirtland's Warbler,* Cranbrook Institute of Science, Bloomfield Hills, Mich., Bull. 40 (1960).

Chapter 8

Page 129. The quotation is from J. T. Curtis, *The Vegetation of Wisconsin,* University of Wisconsin Press, Madison, 1959, p. 234.

Page 132. The peat study was done by G. A. Leisman. See his

paper, "The Rate of Organic Matter Accumulation on the Sedge Mat Zones of Bogs in the Itasca State Park Region of Minnesota," *Ecology,* Vol. 34 (1953), pp. 81–101.

Chapter 9

Page 143. For full details on the bur oak roots, see J. E. Weaver and J. Kramer, "Root System of *Quercus macrocarpa* in Relation to the Invasion of Prairie," *Botanical Gazette,* Vol. 94 (1932), pp. 51–84.

Page 146. The words are those of J. T. Curtis, *op. cit.,* p. 337.

Page 150. H. A. Gleason, "An Isolated Prairie Grove and its Phytogeographical Significance," *Botanical Gazette,* Vol. 53 (1912), pp. 38–49. Also "The Relation of Forest Distribution and Prairie Fires in the Middle West," *Torreya,* Vol. 13 (1913), pp. 173–181.

Page 152. The traverse west from Itasca is described in full in M. F. Buell and J. E. Cantlon, "A Study of Two Forest Stands in Minnesota with an Interpretation of the Forest-Prairie Margin," *Ecology,* Vol. 32 (1951), pp. 194–316; and M. F. Buell and V. Facey, "Forest-Prairie Transition West of Itasca Park, Minnesota," *Bulletin of the Torrey Botanical Club,* Vol. 87 (1960), pp. 46–58.

Chapter 10

Page 155. The first quotation is from W. N. Blane, "A Tour in Southern Illinois in 1822." In M. M. Quaife, *Pictures of Illinois One Hundred Years Ago,* R. R. Donnelly & Sons Co., Chicago, 1918, Part II. The second is from C. Atwater, "On the Prairies and Barrens of the West," *American Journal of Science and Arts,* Vol. 1 (1818), pp. 116–125.

Page 156. The first quotation is from Blane, *op. cit.*

Page 156. Quotations from Hamlin Garland are from *A Son of the Middle Border,* The Macmillan Co., New York, 1920, ch. XIII.

Page 157. Description of the prairie fire is from Blane, *op. cit.*

Page 182. By the late 1960's the Pipestone prairie was growing up to brush. In the light of current knowledge of the relation between fire and prairie, parts of the area were burned in the early 1970's on an experimental basis. The result was a dramatic reversion to the character of a "virgin" prairie, with large, reinvigorated grasses bedecked with reinvigorated prairie flowers.

Chapter 11

Page 185. Quoted from J. S. Olson, "Lake Michigan Dune Development. III. Lake-level, Beach, and Dune Oscillation," *Journal of Geology,* Vol. 66 (1958), pp. 473–483. In the same volume are two more papers of the same series: "I. Wind Velocity Profiles," pp. 254–263; and "II. Plants as Agents and Tools of Geomorphology," pp. 345–351.

Page 190. The study of dune grass seedlings is described by C. C. Laing in "Studies in the Ecology of *Ammophila breviligulata.* I.," *Botanical Gazette,* Vol. 119 (1958), pp. 209–216.

Page 193 and page 196. The relevant paper of Cowles is "The Ecological Relations of the Vegetation on the Sand Dunes of Lake Michigan," *Botanical Gazette,* Vol. 17 (1899), pp. 95–117, 167–202, 281–308, 361–391.

Page 194. Olson's vegetational studies are described in "Rates of Succession and Soil Changes on Southern Lake Michigan Sand Dunes," *Botanical Gazette,* Vol. 119 (1958), pp. 125–170.

Page 198. The history of the Sleeping Bear is given in F. C. Gates, "The Disappearing Sleeping Bear Dune," *Ecology,* Vol. 31 (1950), pp. 386–392. More recent developments are described by W. T. Gillis and K. I. Bakeman, "The Disappearing Sleeping Bear Sand Dune," in *Michigan Botanist,* 1963.

Chapter 12

Page 210. See A. D. Hassler, "Experimental Limnology," *Bio-Science,* Vol. 14, No. 7 (July, 1964), pp. 36–38.

Page 211. See R. H. Hofstrand, "Wild Ricing," *Natural History,* Vol. 79, no. 3 (March, 1970), pp. 50–55.

Page 215. The quotations from Errington are from *Muskrat Populations,* Iowa State University Press, Ames, Iowa, 1963, p. 418 and p. 13.

Page 218. Some of Ridgway's photos are reproduced in A. A. Lindsay et al., "Vegetation and Environment along the Wabash and Tippecanoe Rivers," *Ecological Monographs,* Vol. 31 (1961), pp. 105–156.

Page 220. The experiments on flooded trees were done by Lindsay et al., *op. cit.*

Chapter 13

Histories of the country are, of course, legion. For the Middle West and from the point of view of this book, meaty and readable volumes are:

R. A. Billington, *Westward Expansion: A History of the American Frontier,* The Macmillan Co., New York, 1960.

R. H. Brown, *Historical Geography of the United States,* Harcourt Brace, New York, 1948.

R. C. Buley, *The Old Northwest — Pioneer Period: 1815–1840.* University of Indiana Press, Bloomington, reissued 1963 (original ed. 1950), 2 vols.

Harlan Hatcher, *The Western Reserve: The Story of New Connecticut in Ohio,* Bobbs-Merrill, Indianapolis, 1949.

Walter Havighurst, *Land of Promise: The Story of the Northwest Territory,* The Macmillan Co., New York, 1946.

Index